Lecture Notes in Computer Science 16411

Founding Editors

Gerhard Goos
Juris Hartmanis

Natasha Lepore · Marius George Linguraru
Editors

Low Field Pediatric Brain Magnetic Resonance Image Segmentation and Quality Assurance

Second MICCAI Challenge, LISA 2025
Held in Conjunction with MICCAI 2025
Daejeon, South Korea, September 27, 2025
Proceedings

 Springer

Editors
Natasha Lepore 
Children's Hospital Los Angeles
Los Angeles, CA, USA

Marius George Linguraru
Children's National Hospital
Washington, DC, USA

ISSN 0302-9743   ISSN 1611-3349 (electronic)
Lecture Notes in Computer Science
ISBN 978-3-032-14416-4   ISBN 978-3-032-14417-1 (eBook)
https://doi.org/10.1007/978-3-032-14417-1

Preface

Low-field MRI is becoming an increasingly important tool for expanding access to neuroimaging, particularly in parts of the world where medical and research infrastructure is limited. Its lower cost and portability make it especially well suited for pediatric imaging, where early insights into brain development and pathology can be life-changing. Still, the lower image quality typical of these systems presents challenges for automated quality control and brain segmentation—key steps for ensuring reliable, usable data. As the use of low-field MRI grows in global pediatric neuroimaging, there is a pressing need for improved methods that can meet these challenges and support high-quality research across diverse settings.

After our inaugural year at Medical Image Computing and Computer Assisted Intervention Conference (MICCAI) 2024, we were thrilled to bring back the Low-field pediatric brain magnetic resonance Image Segmentation and quality Assurance (LISA) Challenge for its second edition, held as part of MICCAI in Daejeon, South Korea, from September 22 to 27, 2025. This year's event marked a big leap forward— with over 167 registered participants representing 22 countries across 6 continents, a remarkable 463% increase from 2024's participation.

The LISA Challenge serves as a benchmarking platform for developing and evaluating automatic image analysis and machine learning algorithms. Building on the feedback and results from our 2024 participants, we significantly enhanced both tasks for the 2025 edition. In 2024, many Task 1 participants identified class imbalance issues skewed toward minor to moderate artifacts in the provided dataset. Thus, we strategically expanded our curated dataset by adding 100 new labeled images, each containing at least one artifact type at the highest severity level, bringing the total from 648 to 748 uLF MR images across seven artifact categories, thus providing a more robust and balanced benchmark for algorithm development. For Task 2, we transformed the challenge structure entirely: what began as hippocampal segmentation alone in 2024 evolved into two distinct subtasks in 2025—Task 2a focusing on bilateral hippocampi and Task 2b targeting bilateral caudate and lentiform nuclei. Inspired by the success of multi-label segmentation approaches demonstrated by several 2024 teams, we provided manual segmentations for all three structure types across the same 99 paired high-field (HF) and linearly registered uLF images, empowering participants to explore multi-structure training strategies. We also released additional ventricle segmentations to further support these innovative approaches, though they were not part of the formal evaluation.

Data for both tasks was provided by teams led by Kirsty Donald (University of Cape Town, South Africa), Sadia Parkar, Sidra Kaleem, and Salman Osmani (Aga Khan University, Pakistan), and Victoria Nankabirwa (Makerere University, Uganda).

Fourteen papers were submitted and peer-reviewed using Microsoft's Conference Management Toolkit through a double-blind process, resulting in the acceptance of 11 short papers and an overall acceptance rate of 67%. In these proceedings, we present the work from the 3 top-performing teams for all three tasks, presented as 6 short papers; 3

teams were top performers in all three tasks. An additional 5 short papers are included due to the high quality of their paper submission despite not ranking in the top three of the tasks. The winning team on Task 1 for quality assessment was from Tsinghua University, Beijing, China. The winning team on Task 2a on hippocampal segmentation was from Lausanne University Hospital and University of Lausanne (UNIL), Lausanne, Switzerland. The winning team on Task 2b on basal ganglia segmentation was from the University of Tübingen, Tübingen, Germany.

To promote broader representation of researchers from low- and middle-income countries in neuroimage research, we also partnered with the *Reinforcing Inclusiveness & diverSity and Empowering MICCAI in Low-to-Middle Income Countries* (RISE-MICCAI) summer school, and launched a parallel challenge focused exclusively on Tasks 2a and 2b, and open only to participants of the RISE-MICCAI Summer School 2025. As part of this initiative, we hosted a webinar in July 2025 to introduce the challenge to over 500 summer school students. By the close of the LISA Challenge, nearly 20% of those students had joined as participants. The winning RISE-MICCAI team on both Tasks 2a and 2b was from the University of the Andes, Bogotá, Colombia.

We extend our sincere gratitude to all participating teams, the MICCAI 2025 Conference, and to challenge chairs Lena Maier-Hein and Nicholas Heller for their support, as well as to the Gates Foundation for funding and resources that made the LISA Challenge possible. We also thank our fellow organizers and data contributors whose collective efforts shaped this into a true community endeavor. Alongside the two of us, contributors included Sean Deoni, Jeffrey Tanedo, Rahimeh Rouhi, Austin Tapp, Krithika Iyer, Di Fan, Lauren Lee, Steve Williams, Kirsty Donald, Victoria Nankabirwa, Sadia Parkar, and Salman Osmani. We are additionally grateful to Esther Puyol for her role in extending the challenge's accessibility to participants of the RISE-MICCAI 2025 Summer School. This work was supported by the Gates Foundation under investments INV-047887, INV-087131, INV-005798, INV-018164, INV-004939, and INV-023509, as well as the Wellcome Leap 1kD program (The First 1000 Days; 222076/Z/20/Z).

The 2025 LISA Challenge is built on a growing global effort to make neuroimaging more accessible, equitable, and clinically impactful through ultra-low-field MRI. This year's participants have advanced not only technical innovation, but also representation and collaboration across diverse geographic and resource settings. Thank you for being part of this community—and we'll see you at LISA 2026!

November 2025

Natasha Lepore
Marius George Linguraru

Organization

Program Committee Chairs

Natasha Lepore   Children's Hospital Los Angeles, USA
Marius George Linguraru   Children's National Hospital, USA

Program Committee

Sean Deoni   Bill and Melinda Gates Foundation, USA
Kirsty Donald   University of Cape Town, South Africa
Di Fan   Children's Hospital Los Angeles, USA
Krithika Iyer   Children's National Hospital, USA
Sidra Kaleem   Aga Khan University, Pakistan
Lauren Lee   Children's Hospital Los Angeles, USA
Victoria Nankabirwa   Makerere University, Uganda
Salman Osmani   Aga Khan University, Pakistan
Sadia Parkar   Aga Khan University, Pakistan
Rahimeh Rouhi   Children's Hospital Los Angeles, USA
Austin Tapp   Children's National Hospital, USA
Jeffrey Tanedo   Children's Hospital Los Angeles, USA
Steve Williams   King's College London, UK

Additional Reviewers

Syed Anwar   Children's National Hospital, USA
Xinyang Liu   Children's National Hospital, USA
Abhijeet Parida   Children's National Hospital, USA
Eryn Perry   Children's Hospital Los Angeles, USA
Pooneh Roshanitabrizi   Children's National Hospital, USA

Contents

Tasks 1, 2a and 2b Combined

Task 1 - Automatic Ultra-Low Field MR Image Quality Assessment

BRIQA: Balanced Reweighting in Image Quality Assessment of Pediatric Brain MRI

Alya Almsouti, Ainur Khamitova, Darya Taratynova[✉],
and Mohammad Yaqub

Mohamed bin Zayed University of Artificial Intelligence (MBZUAI), Abu Dhabi, UAE
{Ainur.Khamitova,darya.taratynova}@mbzuai.ac.ae

Abstract. Assessing the severity of artifacts in pediatric brain Magnetic Resonance Imaging (MRI) is critical for diagnostic accuracy, especially in low-field systems where the signal-to-noise ratio is reduced. Manual quality assessment is time-consuming and subjective, motivating the need for robust automated solutions. In this work, we propose BRIQA (Balanced Reweighting in Image Quality Assessment), which addresses class imbalance in artifact severity levels. BRIQA uses gradient-based loss reweighting to dynamically adjust per-class contributions and employs a rotating batching scheme to ensure consistent exposure to underrepresented classes. Through experiments, no single architecture performs best across all artifact types, emphasizing the importance of architectural diversity. The rotating batching configuration improves performance across metrics by promoting balanced learning when combined with cross-entropy loss. BRIQA improves average macro F1 score from 0.659 to 0.706, with notable gains in Noise (0.430), Zipper (0.098), Positioning (0.097), Contrast (0.217), Motion (0.022), and Banding (0.012) artifact severity classification. The code is available at https://github.com/BioMedIA-MBZUAI/BRIQA.

Keywords: Low-field MRI · Quality Assessment · MRI Artifacts · Gradient-Based Reweighting · Class Imbalance

1 Introduction

Brain Magnetic Resonance Imaging (MRI) is an essential imaging modality to study pediatric brain development. In the early postnatal period, the human brain undergoes rapid growth and structural development; therefore, capturing these changes is important for improving our understanding of brain maturation and allowing the early detection of neurodevelopmental conditions [3,4,8,13]. While MRI is considered safe due to the absence of ionizing radiation [3], high-field systems produce loud noise and require children to remain still in enclosed

---

A. Almsouti, A. Khamitova and D. Taratynova—Equal contribution.

© The Author(s) 2026
N. Lepore and M. G. Linguraru (Eds.): LISA 2025, LNCS 16411, pp. 3–14, 2026.
https://doi.org/10.1007/978-3-032-14417-1_1

spaces for extended periods, often needing sedation, which is not ideal. In addition, these systems' high cost and maintenance requirements limit their accessibility in low- and middle-income countries.

To address this, low-field MRIs offer an alternative solution with portable, point-of-care systems, reduced cost, quieter scans, and open designs, eliminating the need for sedation. However, the decreased signal-to-noise ratio in low field MRI poses limitations on the acquired image quality [1], and introduces artifacts, as shown in Fig. 1. This makes image quality assessment essential to ensure that images meet specific standards and support diagnostic reliability, and given that manual Image Quality Assessment (IQA) is time-consuming and costly, automated solutions are crucial. This motivates Task 1 of the LISA Challenge 2025, the automatic assessment of the quality of the MRI scan in seven artifact classes.

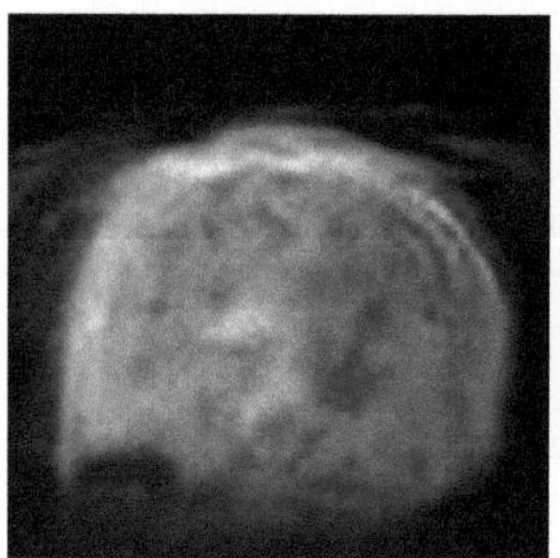 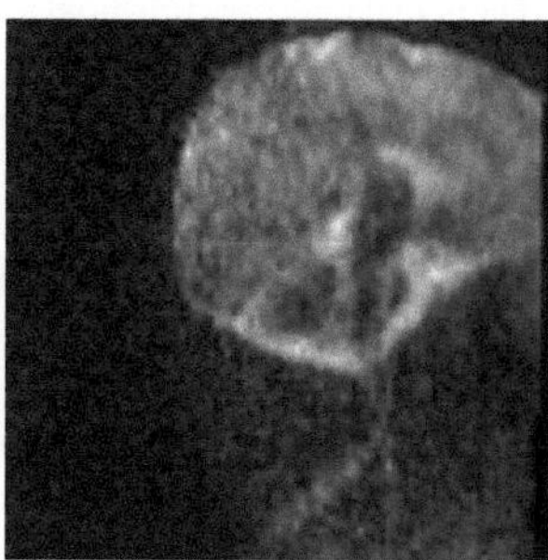 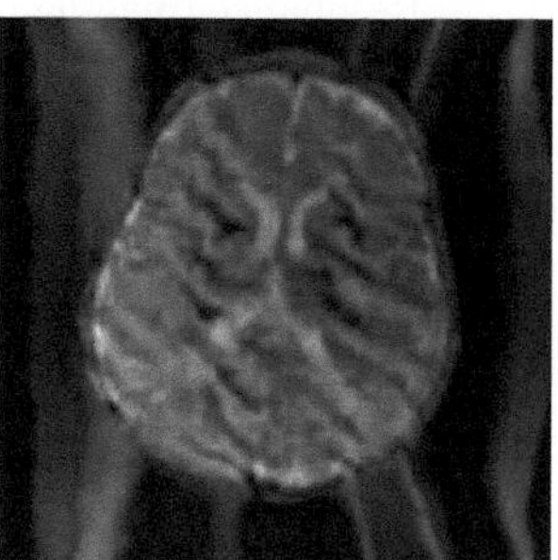

(a) Coronal view: Positioning, Motion, Contrast, and Distortion artifacts.

(b) Sagittal view: Noise, Zipper, and Positioning artifacts.

(c) Axial view: Zipper and Banding artifacts.

Fig. 1. Scans from multiple patients obtained with the 0.064T Hyperfine SWOOP system, showing severe artifacts in different anatomical planes.

Previous efforts have been made in brain MRI IQA, including the machine learning approach by Sanchez et al. [9], which extracts image quality metrics from fetal brain MRI for automatic quality assessment. Deep learning approaches include Zhang et al. [15], who proposed jointly segmenting the brain and assessing quality in fetal MRI slices, while Lou et al. [5] developed a contrastive learning method to enhance feature extraction, leveraging both spatial and frequency representations for quality assessment. The previous LISA 2024 Challenge [4] featured Kim et al. [2], who predicted scan orientation alongside quality assessment, Sundaresan et al. [11], who suggested synthesizing artifacts, and Zhu et al. [16], who developed a multi-label model combining gated CNNs and an ML-Decoder.

However, these prior studies rely on a single model architecture for all artifact types. In practice, performance can vary depending on the medical application;

for example, ResNet may outperform DenseNet in some scenarios and vice versa [6,12]. Moreover, larger models do not necessarily perform better, particularly on small datasets [14]. Therefore, it is beneficial to leverage diverse architectures of varying sizes, as different models may excel at identifying different artifact types based on their distinct visual patterns.

In this work, we introduce BRIQA, a method for the automatic assessment of artifact severity of MRI scans. BRIQA features a tailored model architecture for each artifact type, along with a gradient-weighting strategy and a custom batching technique to address class imbalance. The remainder of this paper is organized as follows: Sect. 2 describes the dataset used and BRIQA framework including gradient-based reweighting and rotating batching, Sects. 3 and 4 presents experimental results with discussion, followed by conclusion at Sect. 5.

2 Methods

2.1 Dataset

In the LISA 2025 Challenge, quality assessment involves scoring the presence of seven common artifacts on pediatric brain magnetic resonance images: Banding, Contrast, Motion, Distortion, Noise, Positioning, and Zipper. Each artifact is rated on a three-point severity scale: 0 for no artifact, 1 for moderate, and 2 for severe. The dataset provided by the challenge organizers consists of 532 brain magnetic resonance images acquired at a low magnetic field strength of 0.064T, representing 244 unique pediatric subjects. Each subject had up to three scans acquired in different orientations: axial, coronal, and sagittal. The severity of artifacts varied across scans.

As illustrated in Table 1, scans containing artifacts are underrepresented. A solution proposed by the first-place winner [11] involved increasing the proportion of scans with artifacts through simulation. Following this approach, we applied artifact simulation using TorchIO, adopting the same parameters as in previous work for all artifact types except motion. Specifically, for moderate motion (level 1), we increased the rotation severity from three to five degrees, and for severe motion (level 2), from seven to ten degrees. These adjustments resulted in more visually distinguishable motion artifacts, ensuring clearer degradation corresponding to the assigned severity level. The distribution of artifacts before and after simulation is shown in Table 1. It is worth noting that although the number of class 1 and 2 instances increased, the overall distribution remains imbalanced. For training, scans were resized to 128×128×128, followed by augmentations such as normalization, center spatial cropping, and random rotation.

2.2 Model Description

To predict the severity of artifacts from MRI scans, we employ a multitask learning framework. As demonstrated by [2], incorporating scan plane classification as an auxiliary task enhances quality assessment, as the appearance of artifacts can vary with anatomical orientation.

Table 1. Distribution of artifact severity before and after simulation.

Artifact	Before			After		
	Class 0	Class 1	Class 2	Class 0	Class 1	Class 2
Noise	426	60	46	734	122	97
Zipper	398	105	29	686	201	66
Positioning	470	47	15	810	106	37
Banding	504	15	13	871	52	30
Motion	384	78	70	672	147	134
Contrast	375	134	23	637	265	51
Distortion	435	56	41	782	101	70

In BRIQA, each input scan $\mathbf{x}$ is processed by an encoder $f_\theta(\cdot)$, which branches into two heads: one for the classification of the severity of the artifact and the other for the classification of the scan plane (axis). To mitigate the effects of class imbalance between severity levels $c' \in 0, 1, 2$, BRIQA adopts a gradient-based loss reweighting strategy.

For each class c, BRIQA calculates how much that class contributes to the training signal by measuring the size of the gradients it produces. Specifically, BRIQA computes the ℓ_2 norm of the gradient of the classification loss $\mathcal{L}_{\mathrm{cls}}^{(c)}$ when considering only the samples that belong to class c. This gradient is taken with respect to the parameters of the classification head, denoted as θ_{cls}. The result is a scalar value ϕ_c, which reflects the overall magnitude of the update that class c would induce on the classification head if it were trained in isolation:

$$\phi_c = \left\| \nabla_{\theta_{\mathrm{cls}}} \mathcal{L}_{\mathrm{cls}}^{(c)} \right\|_2 . \tag{1}$$

To rebalance the contributions from each class, BRIQA normalizes the gradients by the smallest observed norm ℓ_2:

$$\alpha_c = \frac{\min_{c'} \phi_{c'}}{\phi_c}. \tag{2}$$

These weights are then used to compute a weighted classification loss:

$$\mathcal{L}_{\mathrm{cls}} = \sum_{c \in c'} \alpha_c \cdot \mathcal{L}_{\mathrm{cls}}^{(c)}. \tag{3}$$

Finally, the total loss for each batch is defined as:

$$\mathcal{L}_{\mathrm{total}} = \mathcal{L}_{\mathrm{cls}} + \mathcal{L}_{\mathrm{axis}}. \tag{4}$$

where $\mathcal{L}_{\mathrm{axis}}$ encourages orientation-sensitive representations.

Batch Configurations. We experiment with two different configurations to form training batches to handle class imbalance and improve learning stability.

Standard Batching. The first configuration employs standard random sampling, where training batches are formed by shuffling the dataset without enforcing any class distribution constraints.

Rotating Batching. This is a custom configuration that introduces epoch-wise variation in the selection of class 0 samples while maintaining a fixed class ratio within each batch. Each batch of size 4 contains two samples from class 0, one from class 1, and one from class 2. To address the scarcity of class 2, we apply random upsampling to match the number of class 1 samples. Unlike standard random sampling, class 0 examples are drawn using a rotating buffer strategy: their indices are cyclically shifted across epochs using a modular offset. This strategy is illustrated in Fig. 2. This ensures uniform usage of all class 0 samples over time, improving training diversity. Importantly, batches are constructed before data augmentations, enabling consistent per-batch composition while introducing controlled epoch-level variation. To our knowledge, this rotating buffer mechanism is a novel contribution.

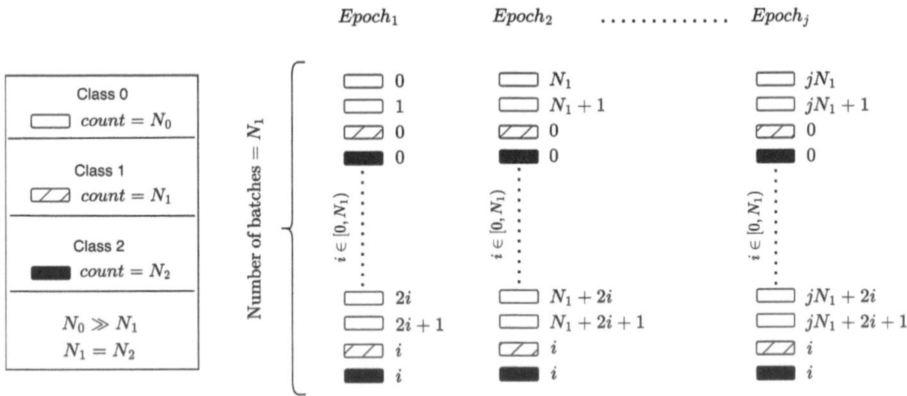

Fig. 2. Rotating batch configuration. Samples from class 0 are drawn using a rotating buffer mechanism, where their indices are cyclically shifted across epochs according to a modular offset. Note that for each sample, we take the remainder of the calculated index with respect to N_0 to achieve the cycling effect; this operation is not shown in the figure.

2.3   Experimental Setup

To evaluate BRIQA, we experimented with a range of encoder backbones, including DenseNet and ResNet variants, as well as MedNeXtS [7]. In addition to backbone comparisons, we compare our method against [16], which considers the problem as multilabel classification. We also compare with the incorporation of a frequency-based encoder from [5]. Specifically, we applied the Discrete Fourier Transform (DFT) to the input MRI and passed the transformed image through a separate encoder. The output of the DFT-based encoder was concatenated with the features of the spatial MRI encoder before classification to capture complementary information in the frequency domain.

To analyze the impact of orientation-aware training, we compared models trained without rotation augmentation to those trained with random rotation up to 180°C. Additionally, we evaluated the effect of different loss functions, including Cross-Entropy (CE), Weighted Cross-Entropy, Focal Loss, and Ordinal Loss [10].

The dataset was split into training and internal validation sets at the patient level, with 80% of patients used for training and 20% for validation. All models were trained for 150 epochs on an NVIDIA A100-SXM4-40GB GPU, using the Adam optimizer with a learning rate of 1×10^{-5}, and a Cosine Annealing learning rate scheduler.

3   Results

Table 2 presents weighted F1 scores for detecting seven MRI artifacts in various encoders with classification heads, showing that no single architecture performs best in all types of artifacts. For example, DenseNet169 excels in detecting Zipper, Positioning, and Noise artifacts (0.844, 0.853, and 0.872, respectively), while Resnet18 achieves the highest scores on Banding, Motion and Contrast (0.947, 0.736, and 0.822, respectively). Simpler models like Resnet10 also perform competitively, particularly on Banding and Noise, outperforming deeper models in some cases. Meanwhile, MedNeXtS [7] struggles with Motion and Distortion. The MLMambaOut [16] architecture, which uses a single backbone for multi-label classification, demonstrates limited effectiveness across several artifact types regardless of the size of the model.

After selecting the best-performing backbone for each artifact, we conducted experiments under different training configurations. Table 3 compares the baseline setup, where reweighting is not applied, with BRIQA. In the baseline, we experiment with various loss functions, regularization techniques, and a fusion model that combines spatial MRI features with Discrete Fourier Transform (DFT) representations.

When comparing loss functions in the baseline setting, we observe that weighted cross-entropy and ordinal loss outperform standard cross-entropy, achieving mean scores of 0.779 and 0.763, respectively, compared to 0.745 without rotation. Interestingly, applying rotation in the cross-entropy setup leads to a 0.016 improvement in the mean score. However, rotation does not consistently yield better performance. For instance, in the regularization setting, the no-rotation variant achieves a mean score of 0.780, the second highest overall, and obtains the best micro-averaged scores, outperforming the rotated version which averages 0.750. Finally, the fusion experiment, which incorporates spectral information via the DFT, shows performance close to that of the second-best configuration, suggesting that frequency-domain features may offer complementary benefits for artifact detection.

Table 2. Weighted F1-scores across different MRI artifact categories using various encoder backbones and MambaOut variants. The best scores per artifact are highlighted in **bold**.

Architecture	Noise	Zipper	Positioning	Banding	Motion	Contrast	Distortion	Mean
Encoder								
Resnet10	**0.844**	0.836	0.826	**0.947**	0.687	0.813	0.765	**0.817**
Resnet18	0.805	0.822	0.830	**0.947**	**0.736**	**0.822**	0.754	**0.817**
Resnet50	0.827	0.803	0.846	**0.947**	0.727	0.800	0.753	0.814
Resnet101	0.815	0.807	0.856	0.926	0.730	0.788	0.753	0.811
DenseNet169	**0.844**	**0.853**	**0.872**	0.926	0.716	0.788	0.776	0.825
DenseNet264	0.698	**0.853**	0.870	0.942	0.731	0.795	0.767	0.808
MedNeXtS [7]	0.807	0.761	0.789	0.881	0.486	0.746	0.601	0.725
MLMambaOut								
MambaOut tiny [16]	0.807	0.845	0.826	0.924	0.722	0.744	0.763	0.805
MambaOut small [16]	0.809	0.838	0.801	0.946	0.690	0.789	**0.821**	0.813
MambaOut base [16]	0.811	0.814	0.834	**0.947**	0.698	0.767	0.790	0.809

Across all configurations, BRIQA improves performance. For cross-entropy without rotation, the mean score increases from 0.745 to the highest overall score of 0.799 with rotating batching. This setup also achieves the best macro scores among all experiments while maintaining weighted and micro scores within 0.01 of the second-best results.

To better understand where these gains are most impactful, Table 4 presents a detailed breakdown of performance improvements in the seven types of MRI artifacts. The model demonstrates consistent improvements in most artifacts, particularly in Noise, Zipper, and Distortion, where all metrics show notable gains. For example, noise and distortion exhibit substantial increases in macro F1 and F2 scores, indicating enhanced sensitivity to rare or harder-to-classify severity levels. Although Banding achieved the highest weighted and micro F1 scores (0.919 and 0.905, respectively), its macro performance was slightly lower than that of other artifacts, probably due to its prevalence in the dataset. Zipper achieved the highest macro F1 score, improving by nearly 10% over baseline.

4   Discussion

Is One Backbone Architecture Enough? Performance in Table 2 suggests that a one-size-fits-all architecture may not be ideal for MRI artifact detection. Instead, leveraging the complementary strengths of diverse backbones could offer improved robustness across artifact types. The consistent variability in per-artifact performance across architectures, especially for more challenging categories like Distortion and Motion, indicates that certain models are more sensitive to specific artifact patterns. Rather than seeking a universally strong

Table 3. Performance metrics for different methods. ○ - No rotation (0°), ↻ - Rotation 180°. Best results are highlighted in **bold**, and second-best results are underlined.

Method	Weighted					Macro					Micro					Mean
	Prec.	Rec.	F1	F2	Acc.	Prec.	Rec.	F1	F2	Acc.	Prec.	Rec.	F1	F2	Acc.	
Baseline																
Standard Batch: Loss Variations																
CE ○	0.800	0.846	0.818	0.834	0.846	0.587	0.552	0.560	0.554	0.552	0.846	0.846	0.846	0.846	0.846	0.745
CE ↻	0.817	0.834	0.818	0.829	0.839	0.708	0.576	0.619	0.590	0.576	0.839	0.840	0.840	0.840	0.839	0.761
Ordinal Loss ○	0.828	0.841	0.821	0.831	0.841	**0.741**	0.572	0.607	0.582	0.572	0.841	0.841	0.841	0.841	0.841	0.763
Weighted CE ○	0.834	0.857	<u>0.842</u>	<u>0.851</u>	0.857	0.661	0.620	0.629	0.622	0.620	<u>0.857</u>	<u>0.857</u>	<u>0.857</u>	<u>0.857</u>	<u>0.857</u>	0.779
Standard Batch: Regularization																
CE ○	0.835	**0.860**	0.842	**0.852**	**0.860**	0.660	0.621	0.628	0.623	0.621	**0.860**	**0.860**	**0.860**	**0.860**	**0.860**	<u>0.780</u>
CE ↻	0.825	0.815	0.810	0.812	0.815	0.670	0.607	0.608	0.605	0.607	0.815	0.815	0.815	0.815	0.815	0.750
Standard Batch: DFT Fusion																
CE ○	0.829	0.845	0.834	0.840	0.845	0.701	0.631	0.659	0.641	0.631	0.845	0.845	0.845	0.845	0.845	0.779
BRIQA																
Standard Batch: Loss Variations																
CE ○	<u>0.840</u>	<u>0.853</u>	0.844	0.849	0.853	<u>0.732</u>	0.657	0.688	0.668	0.657	0.853	0.853	0.853	0.853	0.853	0.794
CE ↻	0.789	0.743	0.763	0.750	0.743	0.576	0.612	0.592	0.607	0.619	0.743	0.743	0.743	0.743	0.743	0.701
Ordinal Loss ○	0.838	0.786	0.801	0.789	0.786	0.653	0.649	0.621	0.625	0.649	0.786	0.786	0.786	0.786	0.786	0.742
Focal loss ○	0.818	0.818	0.818	0.818	0.818	0.642	0.648	0.645	0.647	0.648	0.818	0.818	0.818	0.818	0.818	0.760
Standard Batch: DFT Fusion																
CE ○	0.821	0.827	0.823	0.825	0.827	0.665	0.683	0.671	<u>0.677</u>	0.683	0.827	0.827	0.827	0.827	0.827	0.776
Rotating Batch: BRIQA																
CE ○	**0.843**	0.849	**0.846**	0.848	0.849	0.724	**0.690**	**0.706**	**0.696**	**0.690**	0.849	0.849	0.849	0.849	0.849	**0.799**

backbone, it may be more effective to utilize this diversity and design adaptive frameworks that combine multiple models to capitalize on their respective strengths. We hypothesize that the performance difference comes from how each network propagates features. ResNet adds features through residual connections, capturing global structure and performing better on artifacts affecting the whole image, like motion, contrast, and banding. DenseNet concatenates features, preserving fine detail, which helps with localized or textural artifacts such as zipper lines and positioning shifts.

Is Cross-Entropy Enough? The baseline experiments show that standard cross-entropy loss is suboptimal when compared to both weighted cross-entropy and ordinal loss. These alternative loss functions are more effective in addressing label imbalance and capturing ordinal relationships between classes, resulting in higher macro- and weighted scores. However, when paired with BRIQA, the standard cross-entropy loss achieves the highest overall performance compared to other losses. This improvement comes from the rotating batch configuration's ability to expose the model to diverse artefact combinations across training iterations, helping the model generalise better across underrepresented classes. In addition to weighting the loss based on gradient contributions, which eliminates the need for explicit reweighting mechanisms in focal loss. In this context, stan-

Table 4. Performance metrics of the best-performing model across artifact types. ($\uparrow$) indicates improvement over the CE without gradient reweighting, and ($\downarrow$) denotes a performance decrease.

Metric	Noise	**Zipper**	**Positioning**	**Banding**	**Motion**	**Contrast**	**Distortion**
Weighted							
Precision	$0.863_{\uparrow 0.238}$	$0.871_{\uparrow 0.021}$	$0.902_{\uparrow 0.037}$	$0.936_{\uparrow 0.001}$	$0.755_{\uparrow 0.025}$	$0.811_{\downarrow 0.012}$	$0.870_{\uparrow 0.100}$
Recall	$0.876_{\uparrow 0.086}$	$0.867_{\uparrow 0.010}$	$0.838_{\downarrow 0.048}$	$0.905_{\downarrow 0.057}$	$0.781_{\uparrow 0.029}$	$0.810_{\downarrow 0.019}$	$0.867_{\uparrow 0.057}$
F1-score	$0.865_{\uparrow 0.167}$	$0.863_{\uparrow 0.010}$	$0.860_{\downarrow 0.011}$	$0.919_{\downarrow 0.028}$	$0.747_{\uparrow 0.012}$	$0.799_{\downarrow 0.023}$	$0.861_{\uparrow 0.085}$
F2-score	$0.871_{\uparrow 0.120}$	$0.864_{\uparrow 0.009}$	$0.844_{\downarrow 0.036}$	$0.910_{\downarrow 0.045}$	$0.765_{\uparrow 0.020}$	$0.803_{\downarrow 0.022}$	$0.863_{\uparrow 0.069}$
Accuracy	$0.876_{\uparrow 0.086}$	$0.867_{\uparrow 0.010}$	$0.838_{\downarrow 0.048}$	$0.905_{\downarrow 0.057}$	$0.781_{\uparrow 0.029}$	$0.810_{\downarrow 0.019}$	$0.867_{\uparrow 0.057}$
Macro							
Precision	$0.747_{\uparrow 0.483}$	$0.865_{\uparrow 0.214}$	$0.713_{\uparrow 0.037}$	$0.593_{\downarrow 0.061}$	$0.716_{\uparrow 0.080}$	$0.705_{\downarrow 0.036}$	$0.725_{\uparrow 0.177}$
Recall	$0.728_{\uparrow 0.396}$	$0.684_{\uparrow 0.055}$	$0.794_{\uparrow 0.177}$	$0.643_{\uparrow 0.088}$	$0.598_{\uparrow 0.013}$	$0.738_{\uparrow 0.030}$	$0.636_{\uparrow 0.215}$
F1-score	$0.725_{\uparrow 0.430}$	$0.731_{\uparrow 0.098}$	$0.732_{\uparrow 0.097}$	$0.605_{\uparrow 0.012}$	$0.625_{\uparrow 0.022}$	$0.698_{\downarrow 0.022}$	$0.657_{\uparrow 0.217}$
F2-score	$0.725_{\uparrow 0.408}$	$0.698_{\uparrow 0.068}$	$0.761_{\uparrow 0.138}$	$0.622_{\uparrow 0.053}$	$0.605_{\uparrow 0.014}$	$0.716_{\uparrow 0.004}$	$0.641_{\uparrow 0.216}$
Accuracy	$0.728_{\uparrow 0.395}$	$0.684_{\uparrow 0.055}$	$0.793_{\uparrow 0.177}$	$0.643_{\uparrow 0.088}$	$0.598_{\uparrow 0.013}$	$0.738_{\uparrow 0.030}$	$0.636_{\uparrow 0.215}$
Micro							
Precision	$0.876_{\uparrow 0.086}$	$0.867_{\uparrow 0.010}$	$0.838_{\downarrow 0.048}$	$0.905_{\downarrow 0.057}$	$0.781_{\uparrow 0.029}$	$0.810_{\downarrow 0.019}$	$0.867_{\uparrow 0.057}$
Recall	$0.876_{\uparrow 0.086}$	$0.867_{\uparrow 0.010}$	$0.838_{\downarrow 0.048}$	$0.905_{\downarrow 0.057}$	$0.781_{\uparrow 0.029}$	$0.810_{\downarrow 0.019}$	$0.867_{\uparrow 0.057}$
F1-score	$0.876_{\uparrow 0.086}$	$0.867_{\uparrow 0.010}$	$0.838_{\downarrow 0.048}$	$0.905_{\downarrow 0.057}$	$0.781_{\uparrow 0.029}$	$0.810_{\downarrow 0.019}$	$0.867_{\uparrow 0.057}$
F2-score	$0.876_{\uparrow 0.086}$	$0.867_{\uparrow 0.010}$	$0.838_{\downarrow 0.048}$	$0.905_{\downarrow 0.057}$	$0.781_{\uparrow 0.029}$	$0.810_{\downarrow 0.019}$	$0.867_{\uparrow 0.057}$
Accuracy	$0.876_{\uparrow 0.086}$	$0.867_{\uparrow 0.010}$	$0.838_{\downarrow 0.048}$	$0.905_{\downarrow 0.057}$	$0.781_{\uparrow 0.029}$	$0.810_{\downarrow 0.019}$	$0.867_{\uparrow 0.057}$
Mean	$0.826_{\uparrow 0.216}$	$0.822_{\uparrow 0.040}$	$0.818_{\uparrow 0.019}$	$0.814_{\downarrow 0.020}$	$0.725_{\uparrow 0.027}$	$0.778_{\downarrow 0.012}$	$0.797_{\uparrow 0.113}$

dard cross-entropy benefits from a more uniform and representative training distribution, making it competitive, even surpassing more specialized loss functions.

Does Scan Rotation Always Help? If rotation-based augmentation is applied, the benefits appear inconsistent in different settings. While it yields marginal gains in some configurations at baseline, such as standard cross-entropy, it can degrade performance in more optimized settings like regularized training, where the non-rotated variant achieved a substantially better mean (0.780 vs. 0.750) and the highest micro-average overall. In contrast to baseline, in batch configuration settings, applying rotation with standard cross entropy degraded the performance. This suggests that rotation may inject noise or disrupt spatial integrity in some representations, especially when models already have strong regularisation or batching restrictions.

Can Custom Batching Improve Learning from Imbalanced Data? Our findings highlight the significant impact of batching design on model performance under class imbalance. Unlike standard batching, which randomly samples data and may repeatedly draw from overrepresented classes, rotating batching enforces a fixed class ratio within each batch while systematically cycling through majority class (class 0) samples across epochs. This strategy ensures that minority classes are consistently represented in every batch, while the majority class is varied to maintain diversity and prevent oversaturation. By balancing the gradient signal across classes, rotating batching helps stabilize training and reduces the tendency to overfit to dominant class patterns. This is particularly important in the context of artifact detection, where severe artifact cases (class 2) are underrepresented. Without careful batching, the model could learn to ignore rare artifacts in favor of more frequent, clean scans. Rotating batching ensures that learning remains attentive to all severity levels, improving generalization and classification robustness.

Does Frequency Domain Help? The observed performance of the DFT fusion setup suggests that incorporating frequency-domain features offers complementary benefits to spatial representations. Although the mean score of the DFT fusion model falls slightly below that of the best regularized configuration, its strong performance in macro-averaged metrics indicates improved generalization to underrepresented severity levels. This is particularly relevant for artifacts, where frequency patterns may be more informative than spatial textures alone. However, the relatively modest gains compared to those of other setups suggest that a simple fusion strategy may not fully exploit the potential of spectral information. More advanced integration mechanisms, such as attention-based fusion, may be necessary to effectively combine spatial and frequency-domain features.

Computational Requirements. BRIQA takes 0.008-0.016 s per sample with 740 MB peak GPU memory for DenseNet-based variants (Noise, Zipper, Positioning), 0.021 s and 4.3 GB for ResNet18 models (Banding, Contrast, Motion), and 0.041 s with 7.2 GB for the MedNextS Distortion model.

5   Conclusion

In this work, we addressed the challenge of automatic classification of MRI artifact severity under class imbalance by proposing BRIQA, which integrates axis prediction, gradient-based loss reweighting, and a rotation-based batch construction strategy. Our findings show that architectural diversity can be leveraged for better performance across different artifact categories, while rotating batching significantly enhances generalization by ensuring consistent exposure to minority classes. Future work may explore dynamic ensemble methods based on artifact type or severity distribution, as well as artifact-specific expert models trained to handle visually distinct patterns. While BRIQA demonstrates improved performance, several limitations warrant consideration. First, the multi-architecture approach requires training and maintaining multiple models, increasing computational overhead compared to single-model solutions. Second, the relatively small dataset size and reliance on simulated artifacts may limit generalization to other low-field MRI systems or diverse patient populations. Clinical validation on larger, multi-center datasets is necessary to confirm BRIQA's utility in real diagnostic workflows.

Disclosure of Interests. The authors have no competing interests.

References

1. Arnold, T.C., Freeman, C.W., Litt, B., Stein, J.M.: Low-field mri: clinical promise and challenges. J. Magn. Resonance Imaging **57**(1), 25–44 (2023)
2. Kim, H., Seo, J., Ryu, S., Park, J. H., On, S., Choi, J.: Axis-guided quality assessment and multi-label hippocampal and ventricular segmentation in low-resolution pediatric brain MRI. In: Lepore, N., Linguraru, M.G., eds., Proceedings of the Low Field Pediatric Brain Magnetic Resonance Image Segmentation and Quality Assurance (LISA 2024), vol. 15515 of Lecture Notes in Computer Science, pp. 53–62. Springer, Cham (2025)
3. Lenroot, R.K., Giedd, J.N.: Brain development in children and adolescents: insights from anatomical magnetic resonance imaging. Neurosci. Biobehav. Rev. **30**(6), 718–729 (2006)
4. Lepore, N., Linguraru, M.G.: Low field pediatric brain magnetic resonance image segmentation and quality assurance: first Miccai challenge, LISA 2024, held in conjunction with Miccai 2024, Marrakesh, Morocco, October 10, 2024, Proceedings (2025)
5. Lou, Y., Zhang, J., Xu, D., Cao, Y., Wang, H., Huang, Y.: No-reference MRI quality assessment via contrastive representation: Spatial and frequency domain perspectives. In: 2024 IEEE International Conference on Multimedia and Expo (ICME), pp. 1–6. IEEE (2024)
6. Putra, I.P.G.Y.P., Dewi, N.W.J.K., Lesmana, P.S.W., Suryawan, I.G.T., Putra, P.S.U.: Comparison of RESNET-50 and DENSENET-121 architectures in classifying diabetic retinopathy. Indonesian J. Data Sci., **6**(1), 64–72 (2025)
7. Roy, S., et al.: Mednext: transformer-driven scaling of convnets for medical image segmentation. In: Greenspan, H., et al., (eds.), Medical Image Computing and Computer Assisted Intervention – MICCAI 2023, pp. 405–415, Cham, 2023. Springer Nature Switzerland (2023)
8. Saman, U., Haque, A., Hussain, N., Shamim, B.: Utility of magnetic resonance imaging of brain in neurocritically ill children in pediatric intensive care unit: a single-center retrospective observational study. J. Pediatric Critical Care **11**(1), 6–9 (2024)
9. Sanchez, T., Esteban, O., Gomez, Y., Eixarch, E., Cuadra, M.B.: Fetmrqc: automated quality control for fetal brain MRI, pp. 3–16 (2023)
10. Shi, X., Cao, W., Raschka, S.: Deep neural networks for rank-consistent ordinal regression based on conditional probabilities. Patt. Anal. Appl. **26**(3), 941–955 (2023)
11. Sundaresan, V., Dinsdale, N.K.: Automated quality assessment using appearance-based simulations and hippocampus segmentation on low-field PAEDIATRIC brain MR images, 41–52 (2024)
12. Uemura, T., Näppi, J.J., Hironaka, T., Kim, H., Yoshida, H.: Comparative performance of 3d-DENSENET, 3d-RESNET, and 3d-VGG models in polyp detection for CT colonography. In: Medical Imaging 2020: computer-aided diagnosis, vol. 11314, pp. 736–741. SPIE (2020)
13. Workneh, F., et al.: Feasibility and acceptability of magnetic resonance imaging and electroencephalography for child neurodevelopmental research in rural ethiopia. Front. Public Heal. **13**, 1551982 (2025)
14. Yang, Y., et al.: A comparative analysis of eleven neural networks architectures for small datasets of lung images of covid-19 patients toward improved clinical decisions. Comput. Biol. Med. **139**, 104887 (2021)

15. Zhang, W., et al.: A joint brain extraction and image quality assessment framework for fetal brain mri slices. Neuroimage **290**, 120560 (2024)
16. Zhu, Y., Jiang, H., Cai, R., Chen, G.: Multi-label mambaout for quality assessment of low-field pediatric brain MR images. In: MICCAI Challenge on Low Field Pediatric Brain Magnetic Resonance Image Segmentation and Quality Assurance, pp. 3–11. Springer Nature Switzerland Cham (2024)

Robust Multi-label Classification of MRI Artifacts in Low-Field Neonatal Brain Imaging via View-Conditional Dual-Task Learning

Cristian Lazo-Quispe[1,2]([✉]) and Roberto Espinoza-Chamorro[1]

[1] Graduate School of Informatics, Kyoto University, Kyoto, Japan
`espinozarob@kuhp.kyoto-u.ac.jp`
[2] Universidad Nacional de Ingeniería, Lima, Peru
`cristian.quispe.66r@st.kyoto-u.ac.jp,clazoq@uni.pe`

Abstract. Low-field MRI offers accessible neuroimaging in low-resource settings, but is often degraded by diverse artifacts that compromise diagnostic utility. In this work, we address Task 1 of the LISA Challenge 2025, which involves multi-label ordinal classification of seven artifact types in 3D neonatal brain MRIs acquired at 0.064T. Our pipeline employs a quality-aware 3D-to-2D projection strategy that automatically selects the optimal viewing plane based on voxel resolution, followed by **view-conditional dual-task learning** that jointly predicts artifact severity and brain bounding boxes. By combining brain-focused morphological preprocessing, MaxViT-based feature extraction with view embeddings, and multi-scale probability aggregation across slices, our approach achieves implicit spatial attention without explicit 3D modeling. To handle severe class imbalance and label ambiguity, we apply dynamic focal loss with class-specific weights, stratified patient-level cross-validation, and targeted data augmentation. By aggregating predictions across all valid 2D slices (80–120 per subject), we achieve a weighted F1 score of **0.771** on the test set. We analyze per-artifact performance and demonstrate that efficient 2D view-conditional modeling can match or exceed 3D approaches while maintaining computational efficiency suitable for clinical deployment in resource-limited settings.

Keywords: Low-field MRI · Neonatal Brain Imaging · Multi-label Ordinal Classification · View-Conditional Learning · Dual-Task Learning · MRI Quality Assessment

1 Introduction

Quality assessment of neonatal MRI images, especially from low-field scanners (0.064T), is critical for downstream clinical applications in early brain development and pathology detection [12,13]. Low-field systems offer portability and accessibility in resource-constrained settings [1,2,9], but often suffer from

N. Lepore and M. G. Linguraru (Eds.): LISA 2025, LNCS 16411, pp. 15–25, 2026.
https://doi.org/10.1007/978-3-032-14417-1_2

diverse artifacts that impair image interpretability. Recent advances in portable ultralow-field MRI systems (0.064-0.35T) have demonstrated feasibility for bedside neonatal brain imaging [14], though systematic quality control remains challenging. The LISA Challenge [7] Task 1 focuses on predicting the severity of seven common MRI artifact types: *Noise, Zipper, Positioning, Banding, Motion, Contrast,* and *Distortion.* Each artifact is labeled with an ordinal severity score—0 (Absent), 1 (Mild), or 2 (Severe)—resulting in a multi-label ordinal classification problem with extreme class imbalance and view-dependent annotation.

Prior works on automated MRI quality control, such as MRIQC [3], have focused on adult or high-field datasets and typically rely on hand-crafted image quality metrics. Recent deep learning approaches have demonstrated promise for automated artifact detection [5,10,16], achieving high accuracy on motion, ghosting, and other artifacts in structural MRI. Most of this work addresses binary classification (artifact present/absent) in high-field adult datasets, with limited exploration of ordinal severity grading or low-field pediatric imaging. Deep convolutional networks have proven effective for reference-free quality assessment [4,6], though they remain computationally expensive and sensitive to spacing inconsistencies—particularly problematic in low-field acquisitions where voxel resolution varies significantly across views (1.5mm sagittal vs. 5.0mm axial/coronal in our dataset).

Our Contribution. We propose a view-conditional dual-task learning framework that addresses the unique challenges of low-field neonatal MRI. Our approach automatically selects the highest-resolution viewing plane per subject through quality-aware 3D-to-2D projection, preserving fine anatomical details while maintaining computational efficiency. We integrate learnable view embeddings with FiLM conditioning [8] to capture view-specific artifact patterns, and use auxiliary bounding box prediction to guide the model's spatial attention toward brain regions. This strategy achieved 5th place in the LISA 2025 Challenge (F1=0.771), showing that carefully designed 2D models can match volumetric approaches while remaining practical for resource-limited clinical settings.

2   Methods

Our framework addresses multi-label ordinal artifact classification through quality-aware preprocessing, view-conditional dual-task learning, and probability aggregation. Figure 1 overviews our pipeline. Unlike volumetric approaches requiring extensive 3D convolutions and sensitive to spacing heterogeneity, our method leverages resolution-aware 2D projection for computational efficiency while preserving diagnostic information.

2.1   Dataset Characteristics and Challenges

The LISA Challenge Task 1 dataset comprises 3D brain MRI volumes from neonates scanned at 0.064 T. Each subject has three orthogonal views (sagittal,

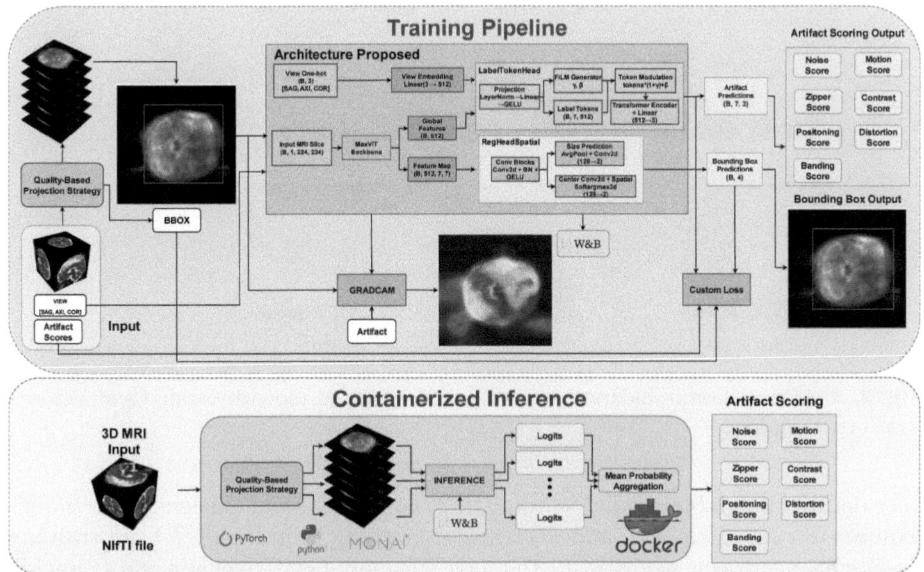

Fig. 1. Training (dual-task learning with spatial attention) and inference (probability aggregation) pipelines.

axial, coronal) with per-view annotations for seven artifact types at three severity levels (0=Absent, 1=Mild, 2=Severe).

Extreme Class Imbalance. Figure 2 shows severe class imbalance with >96% samples in class 0 for *Banding* and *Positioning*, while severe artifacts (class 2) constitute <5% across all categories, necessitating specialized training strategies.

Heterogeneous Voxel Spacing. The dataset exhibits bimodal voxel spacing: approximately 44% of volumes have 1.5mm spacing (predominantly sagittal: X=231, Z=241 subjects), 33% have 5.0mm spacing (predominantly axial/coronal: X=175, Y=181, Z=175 subjects), and 23% have intermediate 1.6mm spacing (X=98, Y=128, Z=98 subjects). This 3.3× resolution difference between high and low-resolution views challenges 3D CNNs that assume spatial consistency. Our quality-aware projection strategy addresses this by automatically selecting the highest-resolution axis per subject, preserving fine anatomical details while maintaining computational efficiency.

2.2   Quality-Aware 3D-to-2D Projection

We propose quality-aware projection that selects the highest-resolution view per subject, motivated by: (1) resolution preservation (1.5mm sagittal captures

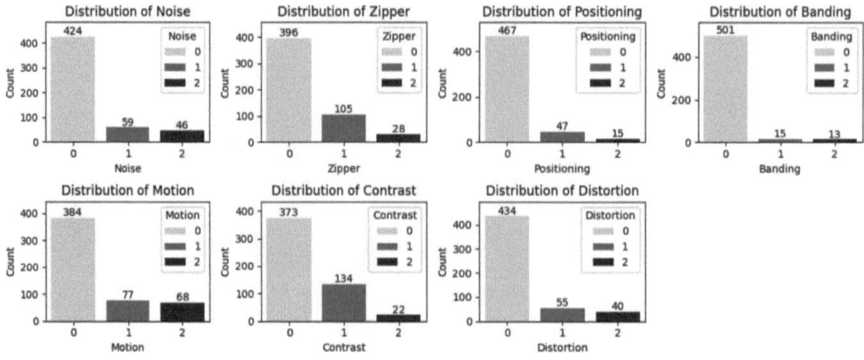

Fig. 2. Extreme class imbalance across 7 artifact types motivates our focal loss and stratified sampling.

finer details than 5.0mm axial/coronal), (2) computational efficiency (2D models require $\sim$8$\times$ less memory), and (3) transfer learning from ImageNet pre-training.

For each subject, we identify the axis with smallest voxel spacing (typically sagittal) and extract all slices from that orientation. Each slice is resized to 256$\times$256 pixels using bilinear interpolation. We apply brain-focused morphological preprocessing (Sect. 2.3) to each slice individually, discarding slices with less than 10% brain coverage. This filtering typically retains 80–120 slices per subject (from an initial 150–200), removing non-informative background slices while preserving all anatomically relevant regions. During training, each slice serves as an independent training sample for the dual-task classifier, allowing the model to learn artifact patterns across varying anatomical contexts within the same subject.

2.3   Brain-Focused Morphological Preprocessing

Low-field MRI suffers from background signals and edge artifacts. We apply multi-stage brain localization to each 2D slice independently: (1) intensity threshold ($I \geq 0.15 \cdot I_{\max}$) for initial mask M_0, (2) morphological refinement via erosion$\rightarrow$dilation$\rightarrow$opening with $k = 3$ kernels, (3) largest connected component retention, and (4) slice filtering—we discard slices where the brain mask covers less than 10% of the image area. This aggressive filtering removes approximately 40–50% of slices per subject (primarily background slices at volume extremes), substantially reducing training time while retaining all diagnostically relevant anatomy. Figure 3 illustrates consistent brain isolation across different slice positions and views.

2.4   View-Conditional Dual-Task Architecture

Our dual-task network jointly learns artifact classification and spatial localization, with bounding box prediction providing implicit attention to brain regions.

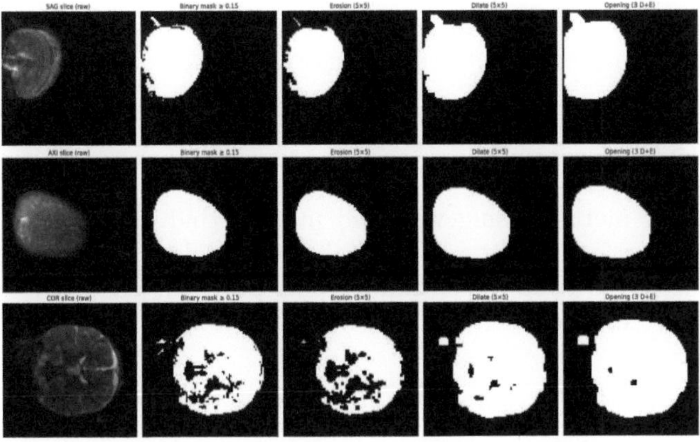

Fig. 3. Brain localization via thresholding and morphological operations across three views.

Backbone and View Conditioning. We use MaxViT-Tiny [15] pre-trained on ImageNet-1K. Learnable view embeddings $\mathbf{e}_v$ for sagittal/axial/coronal are integrated via FiLM [8]: $\mathbf{f}'_i = \gamma_v \odot \mathbf{f}_i + \beta_v$, where $\gamma_v, \beta_v = \mathrm{MLP}(\mathbf{e}_v)$, enabling view-specific artifact pattern learning. The choice of this model was motivated by its robustness and lightweight design, which enabled fast training and inference during the experiments.

Dual-Task Heads. (1) *Classification Head*: 7 learnable artifact tokens with cross-attention yield $\mathbf{P} \in \mathbb{R}^{7 \times 3}$ predictions. (2) *Spatial Head*: predicts bounding box $\mathbf{b} \in \mathbb{R}^4$ to encourage spatial encoding.

Loss Function. We optimize: $\mathcal{L}_{\text{total}} = \mathcal{L}_{\text{focal}} + 0.2(\mathcal{L}_{\text{L1}} + \mathcal{L}_{\text{GIoU}})$, where focal loss uses $\alpha_c \in \{0.25, 0.75, 1.0\}$ for classes $\{0, 1, 2\}$ and $\gamma = 2.0$.

Training. We use 5-fold stratified cross-validation, stratifying by patient ID and artifact labels to ensure balanced folds. Each 2D slice (after morphological filtering) serves as an independent training sample, effectively multiplying the dataset size—a typical subject contributes 80–120 training samples. Class-weighted sampling ($\propto 1/\sqrt{n_c}$) addresses severe imbalance without overwhelming the dominant class. Our augmentation pipeline includes gamma correction, random rotations ($\pm 15°$), elastic deformations, and MixUp [17] ($\alpha = 0.4$). We train with AdamW optimizer (lr=3×10^{-4}, weight decay=10^{-4}), cosine annealing with ReduceL-ROnPlateau, and early stopping after 20 epochs without improvement.

Inference. For each test volume, we extract all slices from the highest-resolution axis following the same preprocessing pipeline. Each slice generates independent predictions from the dual-task classifier. We aggregate per-slice probability predictions through simple averaging: $\mathbf{P}_{\text{final}} = \frac{1}{S} \sum_{s=1}^{S} \mathbf{P}^{(s)}$, where S is the number of valid slices (typically 80–120 per subject). This multi-slice ensemble yields +3.2% F1-macro improvement over using only the central slice.

3   Results

We evaluate our method using weighted metrics that account for class imbalance, reporting performance on both local cross-validation and the official LISA Challenge test set. Our approach ranked **5th place** in Task 1 with a weighted F1 score of 0.771 on the final test phase.

3.1   Overall Performance

Table 1 presents the complete evaluation across local validation and the official test set. Our view-conditional dual-task model with probability aggregation demonstrates strong and consistent performance across all metrics.

Table 1. Performance on local cross-validation and official test set.

Phase	F1-macro	F2-macro	Precision	Recall	Accuracy	Avg
Local CV	0.691	0.577	0.692	0.834	0.834	0.726
Test Set	**0.771**	**0.776**	**0.773**	**0.783**	**0.783**	**0.777**

The test set performance (weighted average: 0.777) demonstrates effective generalization with balanced precision (0.773) and recall (0.783). The +8.0% F1 improvement from local CV to test set suggests our preprocessing and augmentation strategies successfully prevent overfitting despite extreme class imbalance.

3.2   Ablation Study

Table 2 shows progressive improvements from adding architectural components. View conditioning provides the largest gain (+5.2% F1-macro), confirming view-dependent artifact patterns. Adding the auxiliary bounding box task improves performance by +2.5% through implicit spatial supervision. Finally, aggregating predictions across all valid slices per subject (rather than using only the central slice) yields an additional +3.2% improvement.

3.3   Per-Artifact Analysis

Table 3 shows performance varies substantially across artifact types. The model excels at detecting Noise (F1-macro=0.797), Zipper (0.709), and Distortion (0.754), which exhibit distinctive spatial patterns well-captured by MaxViT's multi-scale attention. Positioning (0.432) and Contrast (0.443) prove most challenging due to extreme imbalance and ambiguous visual signatures. Positioning

Table 2. Ablation study showing impact of each component.

Configuration	F1-macro	F1-micro	F2-macro	F2-micro
Baseline (no conditioning)	0.582	0.791	0.498	0.791
+ View conditioning	0.634	0.815	0.549	0.815
+ Bounding box task	0.659	0.826	0.568	0.826
+ Probability aggregation	**0.691**	**0.834**	**0.577**	**0.834**

may be inherently difficult as it relates to patient placement rather than clear image degradation. Contrast variations are global properties potentially requiring whole-volume analysis rather than slice-level assessment. The consistently high F1-micro (0.745âĂŞ0.954) reflects accurate classification of dominant class 0, while lower F2 scores (0.365âĂŞ0.725) indicate missed positive cases despite class balancing strategies.

Table 3. Per-artifact performance on local cross-validation.

Artifact	F1-macro	F1-micro	F2	Dist. (0/1/2)
Noise	**0.797**	0.817	0.702	424 / 59 / 46
Zipper	**0.709**	0.824	0.725	396 / 105 / 28
Distortion	**0.754**	0.804	0.564	434 / 55 / 40
Motion	0.594	0.869	0.617	384 / 77 / 68
Banding	0.596	0.954	0.588	501 / 15 / 13
Contrast	0.443	0.745	0.479	373 / 134 / 22
Positioning	0.432	0.824	0.365	467 / 47 / 15
Average	**0.691**	**0.834**	**0.577**	-

3.4 Qualitative Analysis

Figure 4 shows GradCAM [11] activation maps for three artifact types. The model focuses primarily on brain parenchyma rather than background, validating our dual-task design where auxiliary bounding box prediction provides implicit spatial supervision. Positioning artifacts show broadly distributed attention capturing global spatial relationships, Noise artifacts display activation across the entire brain reflecting global signal degradation, while Zipper artifacts exhibit localized patterns corresponding to phase-encoding directions. This variation in attention patterns suggests artifact-specific spatial priors explaining the differential performance in Table 3.

Figure 5 illustrates potential annotation inconsistencies. Three samples labeled Distortion=2 (severe) show varying visual quality: samples (a-b) display

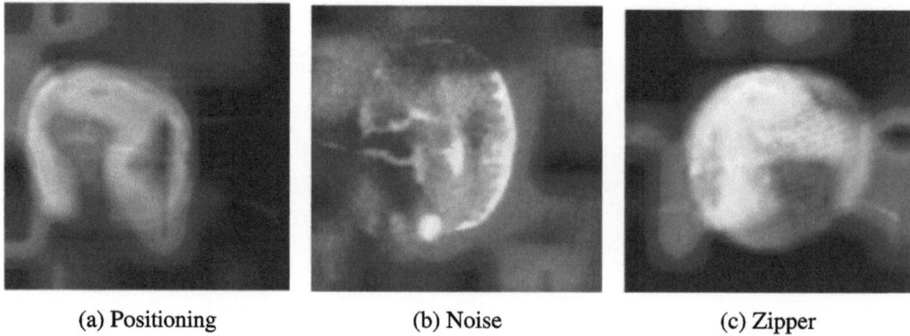

(a) Positioning   (b) Noise   (c) Zipper

Fig. 4. GradCAM activation maps showing spatial attention patterns. Red indicates high activation, blue low activation. (Color figure online)

obvious geometric distortion and intensity artifacts, while (c) appears anatomically intact. This suggests view-dependent artifacts where distortion exists in axial/coronal but not sagittal views, or annotation inconsistencies in severity assessment. Such cases highlight the importance of multi-view consistency checks and motivate future multi-view ensemble approaches integrating information across orthogonal planes.

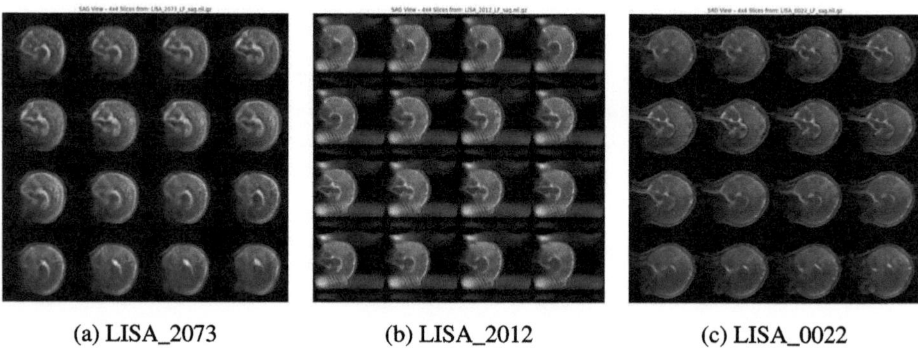

(a) LISA_2073   (b) LISA_2012   (c) LISA_0022

Fig. 5. Three sagittal views labeled Distortion=2. (a-b) show clear artifacts, (c) appears clean, suggesting annotation inconsistency.

4   Discussion

Our view-conditional dual-task learning framework achieves competitive performance (5th place, weighted F1=0.771) while maintaining efficiency suitable for resource-limited clinical settings. The ablation studies reveal that view conditioning provides the largest gain (+5.2% F1-macro), confirming artifact patterns are view-dependent. The auxiliary bounding box task adds $+2.5\%$ through implicit spatial attention, as evidenced by GradCAM visualizations focusing on brain parenchyma rather than background. Multi-slice aggregation contributes $+3.2\%$ by averaging predictions across all valid slices per subject (80–120 slices typically).

Our slice-level approach, where each 2D slice is processed independently, maximizes training data utilization—a typical volume yields 80–120 training samples after filtering. This substantially exceeds composite approaches and allows the model to learn artifact patterns across varying anatomical contexts (basal ganglia, ventricles, cortex) within the same subject. Variable numbers of slices per subject pose no architectural constraints since aggregation occurs only at inference time.

Performance on rare classes remains challenging despite focal loss and class-weighted sampling. Positioning (F1=0.432) and Contrast (F1=0.443) suffer from limited positive examples (15 and 22 severe cases respectively) and ambiguous visual signatures that may require whole-volume analysis. The annotation inconsistency case study highlights that view-dependent artifacts pose fundamental challenges when labels are assigned per-subject rather than per-view, suggesting need for multi-view consistency checks during annotation.

5   Future Work and Conclusion

Future Directions. Multi-view ensemble architectures with cross-view attention could address blind spots inherent to single-plane approaches—our annotation inconsistency analysis (Fig. 5) suggests artifacts may manifest differently across orientations. Ordinal regression losses that explicitly model severity ordering constraints might better capture the continuum from absent to severe artifacts. For highly imbalanced classes like Positioning (F1=0.432), generative oversampling via diffusion models could augment training data while preserving anatomically plausible variations. Foundation model fine-tuning on larger multi-site datasets and uncertainty quantification for clinical deployment represent additional promising directions.

Broader Impact. This work addresses a critical gap in automated quality control for low-field neonatal MRI—a technology poised to transform neuroimaging access in resource-limited settings, yet hindered by frequent artifacts that waste precious scan time with vulnerable patients. Our view-conditional dual-task learning framework demonstrates that carefully designed 2D approaches can achieve competitive performance (5th place, F1=0.771) while maintaining the efficiency necessary for real-world deployment.

Our analysis reveals fundamental challenges beyond algorithm design. The annotation inconsistencies we observed (Fig. 5) suggest current per-subject labeling may be insufficient for view-dependent artifacts, motivating more granular annotation protocols in future datasets. Our ablation studies confirm that incorporating domain knowledge—through quality-aware projection and morphological preprocessing—remains crucial even with modern architectures.

The containerized pipeline (<0.2 s inference) enables real-time feedback during acquisition, allowing technicians to adjust protocols immediately rather than discovering quality issues post-scan. For low-field systems deployed at bedside in neonatal intensive care units, where scan windows are limited and patient repositioning is challenging, this capability could substantially improve clinical workflow efficiency. We make our complete implementation publicly available at https://github.com/CristianLazoQuispe/lisa-challenge2025-task1 to accelerate adoption in low-resource clinical environments worldwide.

References

1. Arnold, T.C., Freeman, C.W., Litt, B., Stein, J.M.: Low-field MRI: clinical promise and challenges. J. Magn. Reson. Imaging **57**(1), 25–44 (2023)
2. Cawley, P., et al.: Development of neonatal-specific sequences for portable ultralow field magnetic resonance brain imaging: a prospective, single-centre, cohort study. EClinicalMedicine **65** (2023)
3. Esteban, O., Birman, D., Schaer, M., Koyejo, O.O., Poldrack, R.A., Gorgolewski, K.J.: Mriqc: Advancing the automatic prediction of image quality in MRI from unseen sites. PLoS ONE **12**(9), e0184661 (2017)
4. Fantini, I., Yasuda, C., Bento, M., Rittner, L., Cendes, F., Lotufo, R.: Automatic MR image quality evaluation using a deep CNN: A reference-free method to rate motion artifacts in neuroimaging. Comput. Med. Imaging Graph. **90**, 101897 (2021)
5. Garcia, M., Dosenbach, N., Kelly, C.: Brainqcnet: a deep learning attention-based model for the automated detection of artifacts in brain structural MRI scans. Imaging Neurosci. **2**, 1–16 (2024)
6. Jimeno, M.M., Ravi, K.S., Fung, M., Vaughan Jr, J.T., Geethanath, S.: Automated detection of motion artifacts in brain MR images using deep learning and explainable artificial intelligence. arXiv preprint arXiv:2402.08749 (2024)
7. LISA Challenge 2025 Organizers: LISA Challenge 2025: Low-field MRI Image Quality Assessment (2025). https://www.synapse.org/Synapse:syn65670170/wiki/631438. Accessed: August 4, 2025
8. Perez, E., Strub, F., De Vries, H., Dumoulin, V., Courville, A.: Film: Visual reasoning with a general conditioning layer. In: Proceedings of the AAAI Conference on Artificial Intelligence. vol. 32 (2018)
9. Raad, J.D., Chinnam, R.B., Arslanturk, S., Tan, S., Jeong, J.W., Mody, S.: Unsupervised abnormality detection in neonatal MRI brain scans using deep learning. Sci. Rep. **13**(1), 11489 (2023)
10. Samani, Z.R., Alappatt, J.A., Parker, D., Ismail, A.A.O., Verma, R.: Qc-automator: Deep learning-based automated quality control for diffusion MR images. Front. Neurosci. **13**, 1456 (2020)
11. Selvaraju, R.R., Cogswell, M., Das, A., Vedantam, R., Parikh, D., Batra, D.: Gradcam: Visual explanations from deep networks via gradient-based localization. In: Proceedings of the IEEE International Conference on Computer Vision, pp. 618–626 (2017)
12. Shen, D.D., et al.: An automatic and accurate deep learning-based neuroimaging pipeline for the neonatal brain. Pediatr. Radiol. **53**(8), 1685–1697 (2023)
13. Sien, M.E., et al.: Feasibility of and experience using a portable MRI scanner in the neonatal intensive care unit. Arch. Dis. Child. Fetal Neonatal Ed. **108**(1), 45–50 (2023)

14. Sun, Z., et al.: A low-field MRI dataset for spatiotemporal analysis of developing brain. Sci. Data **12**(1), 109 (2025)
15. Tu, Z., et al.: Maxvit: Multi-axis vision transformer. In: European Conference on Computer Vision, pp. 459–479. Springer (2022)
16. Vakli, P., et al.: Automatic brain MRI motion artifact detection based on end-to-end deep learning is similarly effective as traditional machine learning trained on image quality metrics. Med. Image Anal. **88**, 102850 (2023)
17. Zhang, H., Cisse, M., Dauphin, Y.N., Lopez-Paz, D.: mixup: Beyond empirical risk minimization. arXiv preprint arXiv:1710.09412 (2017)

Task 2b - Automatic Basal Ganglia Segmentation from Ultra-Low Field MRI

Towards Robust Basal Ganglia Segmentation in Ultra-Low-Field Pediatric MRI via an Optimized MS-TCNet

Yi Liu, Yueyue Zhu, Haotian Jiang, Xiaoyu Bai, Rongqing Cai,
and Geng Chen$^{(\boxtimes)}$

National Engineering Laboratory for Integrated Aero-Space-Ground-Ocean Big Data
Application Technology, School of Computer Science and Engineering, Northwestern
Polytechnical University, Xi'an, China
`geng.chen@ieee.org`

Abstract. Magnetic Resonance Imaging (MRI) provides a non-invasive means to examine pediatric brain anatomy. However, in low-resource settings, ultra-low-field scanners are more widely used due to their affordability and portability, but they often generate images with a low signal-to-noise ratio and poor tissue contrast. This severely hampers the accurate delineation of critical subcortical structures such as the basal ganglia. In this work, we adapt and refine the multi-scale Transformer – CNN network (MS-TCNet) for bilateral basal ganglia segmentation in 0.064T pediatric MRI. Our optimized version, OMS-TCNet, integrates improved data augmentation and fine-tuned training configurations. Extensive experiments on the challenge dataset demonstrate that our method achieves robust and reliable segmentation performance under ultra-low-field imaging conditions. The code is publicly available at: https://github.com/Onion-Boy/OMS-TCNet.

Keywords: Basal Ganglia Segmentation · Low-Field MRI · Pediatric Brain

1 Introduction

Magnetic Resonance Imaging (MRI) offers non-invasive visualization of brain anatomy without ionizing radiation, making it an essential tool in both research and clinical practice. Conventional High-Field (1.5T or 3T) MRI systems impose substantial costs, maintenance demands, and infrastructural requirements that place them out of reach for many low- to middle-resource countries. In contrast, Low-Field (LF) MRI scanners operating below 0.5T have gained popularity due to their lower cost and simplified installation requirements. Among them, Ultra-Low-Field (ULF) systems below 0.1T offer even greater portability and affordability, along with the added benefit of imaging children without the need for

---

Y. Liu and Y. Zhu—Equal contribution.

N. Lepore and M. G. Linguraru (Eds.): LISA 2025, LNCS 16411, pp. 29–38, 2026.
https://doi.org/10.1007/978-3-032-14417-1_3

sedation [5,8]. However, the markedly reduced signal-to-noise ratio and tissue contrast in LF and ULF imaging often lead to severe artifacts such as noise, motion blur, and indistinct boundaries, which pose substantial challenges for the automated analysis of critical brain structures.

Among these structures, the basal ganglia play fundamental roles in motor control, cognition, and behavior regulation. This group of nuclei mainly includes the caudate nucleus and lentiform nucleus. Morphological and volumetric alterations in the basal ganglia have been linked to a range of neurological and psychiatric disorders, such as schizophrenia, Alzheimer's disease, and others. Accurate segmentation of these nuclei is essential for both diagnostic evaluation and the planning of stereotactic neurosurgical procedures [9]. However, segmenting the basal ganglia presents significant challenges. Manual delineation is time-consuming and prone to substantial inter-rater variability. These difficulties are further exacerbated in LF and ULF scans due to low tissue contrast and pronounced class imbalance between adjacent subcortical nuclei [6].

Historically, automatic basal ganglia segmentation methods have fallen into four main categories [14]: atlas-based registration [1], statistical shape modeling [10], deformable surface frameworks [15] and deep learning-based approaches [7]. The first three methods leverage prior anatomical knowledge or structural constraints, yet they often struggle to generalize across individuals, particularly in pediatric populations where brain morphology changes rapidly. These traditional techniques are also highly sensitive to the poor tissue contrast and artifacts characteristic of ULF MRI. More recently, deep learning-based methods such as CNN-based networks, exemplified by U-Net [12] and its variants employing encoder – decoder architectures with skip connections, as well as Transformer-based architectures [3,4] and state-space models like Mamba [16,17], have shown great promise in medical image analysis. However, their performance degrades substantially under ULF conditions due to reduced contrast, high noise, and limited training data, which together make accurate modeling of fine anatomical details particularly challenging.

To address the challenges of segmenting basal ganglia structures in ULF pediatric brain MRI, we adopt and optimize the MS-TCNet architecture [2], a hybrid Transformer – CNN combined network learning multi-scale features. While MS-TCNet has demonstrated effectiveness in segmenting brain tumors, its direct application to ULF pediatric MRI scans has not been explored, which are characterized by extreme noise, low contrast, and small anatomical structures. To fill this gap, we present an optimized version of the model, termed Optimized MS-TCNet (OMS-TCNet), tailored specifically for the automatic segmentation of the bilateral basal ganglia in 0.064T pediatric T2-weighted MRI. OMS-TCNet retains the core architectural components of MS-TCNet, including a Transformer-based encoder for extracting multi-scale global features, a CNN-based decoder for hierarchical refinement, and a Multi-Scale Feature Fusion (MSFF) module to aggregate predictions across resolution levels. On top of this baseline, we make the following key contributions:

1. We are the first to adapt MS-TCNet for bilateral basal ganglia segmentation in 0.064T pediatric MRI, verifying its applicability in ULF conditions.
2. We design a diverse augmentation pipeline using the batchgenerators library to enhance generalizability and robustness.
3. We optimize training performance by systematically tuning key configurations, including input patch size and batch size.
4. We validate OMS-TCNet on the challenge dataset, achieving an average Dice score of 0.85 ± 0.05 and demonstrating accurate, reliable segmentation performance.

The remainder of this paper is organized as follows: Sect. 2 presents our methodological pipeline, including a recall of the MS-TCNet architecture, data augmentation strategies, and training configuration optimization. Section 3 provides both quantitative and qualitative evaluation results. Section 4 offers a brief discussion of the findings and limitations, and Sect. 5 concludes the paper.

2   Method

2.1   Overview of OMS-TCNet

To tackle the challenge of segmenting the bilateral basal ganglia from ULF (0.064T) pediatric brain MR images, we adopt and customize the publicly available implementation of MS-TCNet, which is originally proposed for other medical image segmentation tasks. We first reproduce the baseline model to ensure correctness, then make a series of optimizations to better adapt it to the unique characteristics of ULF pediatric MRI.

Our improvements span three main aspects. First, we enhance the training data with a carefully designed augmentation pipeline to improve generalization under data scarcity and high noise conditions. Second, we conduct a set of experiments to tune critical hyperparameters, including patch size and batch size, which prove essential for improving segmentation accuracy. Finally, while retaining the original model architecture, we integrate these changes into an optimized version, which we refer to as OMS-TCNet.

2.2   MS-TCNet Architecture Recall

MS-TCNet is a multi-scale hybrid segmentation network that integrates Transformer-based global feature extraction with CNN-based local refinement. As shown in Fig. 1, MS-TCNet follows an encoder – decoder structure, where the encoder captures hierarchical features using shunted Transformer blocks, and the decoder progressively refines these features through a multi-scale CNN-based decoder.

The encoder begins with a patch embedding module that uses stacked 3D convolutions to convert the input volume into an initial low-resolution feature map. This is followed by four encoding stages, each composed of a linear downsampling block and two Shunted Transformer Blocks. Within each Transformer Block, the

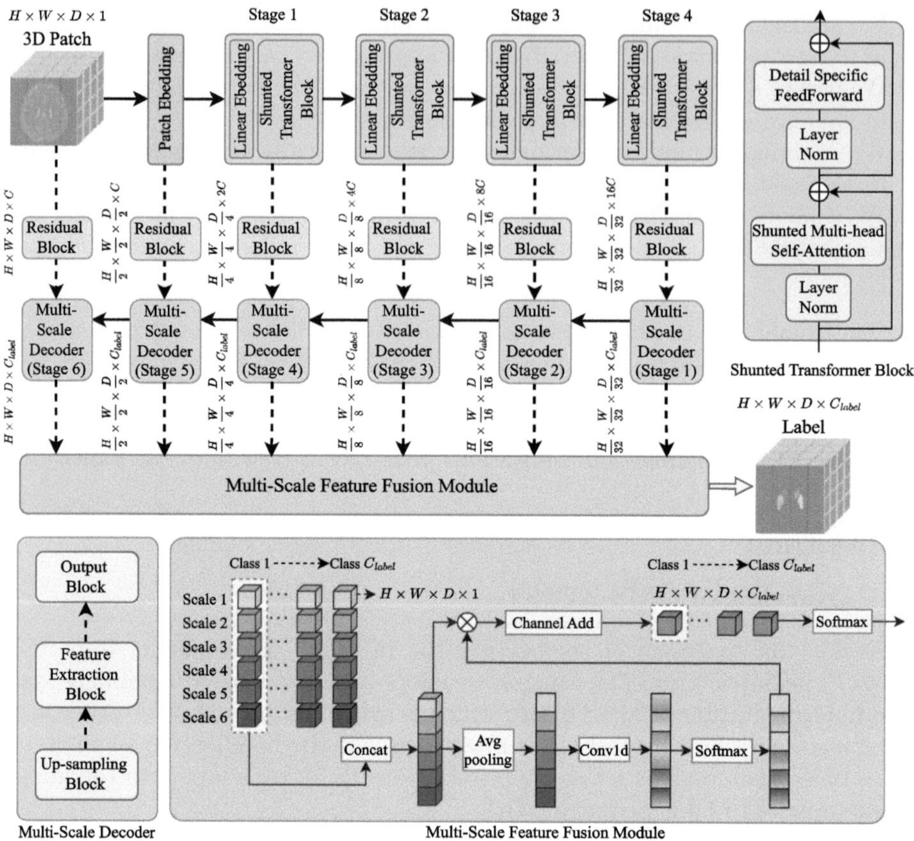

Fig. 1. Overall architecture of MS-TCNet, comprising a Transformer-based encoder, a CNN-based decoder, and a multi-scale feature fusion module (illustration redrawn based on [2]).

Shunted Multi-Head Self-Attention (SMSA) [11] mechanism introduces diverse receptive fields across heads by applying different spatial downsampling rates. This allows the network to model both fine and coarse contextual dependencies. Each block also includes a Detail-Specific Feed-Forward Network (DSFF) [11] to further enhance local structural information.

The decoder consists of six sequential stages, each decoding features at a distinct resolution scale to progressively refine semantic details across multiple levels. At each stage, the previous feature map is upsampled and fused with the corresponding encoder output via a Residual Block. Feature Extraction (FE) blocks containing residual units and SCSE attention modules [13] are applied to strengthen meaningful features while suppressing irrelevant ones. Finally, an output block generates scale-specific segmentation maps, which are combined in the Multi-Scale Feature Fusion (MSFF) module.

The MSFF module first upsamples all decoder outputs to the original image resolution, then aggregates per-class multi-scale predictions using an attention-like mechanism. Global average pooling and a 1D convolution are applied across the scales to compute adaptive weights, which are used to generate a weighted sum of features for the final prediction.

2.3 Data-Level Optimization: Augmentation Strategy

To mitigate overfitting and enhance the generalization capability of the model, we implement a comprehensive data augmentation pipeline using the batchgenerators framework. This open-source library provides widely used augmentation utilities for 3D medical image segmentation and allows flexible customization of transformation parameters. The pipeline introduces diverse appearance and spatial variations that simulate realistic perturbations in pediatric ULF MRI scans.

Most of the employed data augmentation strategies follow standard practices commonly used in 3D medical image segmentation, with several parameters slightly adjusted to accommodate the characteristics of ultra-low-field MRI. Specifically, the augmentation pipeline begins with 3D spatial transformations to promote geometric variability. Each patch undergoes random rotations along the x, y, and z axes, with angles sampled uniformly from ± 30 degrees. Isotropic scaling is also applied, with scale factors sampled from the range 0.7 to 1.4. These spatial transformations are applied with a probability of 0.2 per sample and a per-axis rotation probability of 1.0, which were intentionally reduced to prevent excessive geometric deformation given the small size of the basal ganglia structures in ultra-low-field images.

To enhance robustness against noise and imaging artifacts common in ULF MRI, we incorporate several intensity-based augmentations. Gaussian noise is added with a probability of 0.1 per sample, while Gaussian blur is applied using σ sampled from [0.5, 1.0], with a probability of 0.2 per sample and 0.5 per channel. Brightness and contrast perturbations are each applied with a probability of 0.15 per sample to simulate variability in image acquisition. Gamma correction is performed in two modes: one with histogram normalization (range $0.7 - 1.5$, probability 0.1) and one without (same range, probability 0.3).

Mirror flipping along the axial, coronal, and sagittal planes ($[0, 1, 2]$) is also employed to introduce left – right anatomical symmetry into the training data. For consistency, label values of -1 are reassigned to 0 prior to training. Finally, both image and label arrays are converted into PyTorch tensors. All transformations are composed into a single pipeline using Compose, ensuring efficient on-the-fly augmentation during data loading.

In order to assess the individual impact of augmentation components, we design and evaluate four augmentation settings: full augmentation, spatial-only, mirror-only, and no augmentation. This comparison helps identify which transformations contribute most to model robustness in the context of noisy, low-contrast brain MRI. Among all configurations, full augmentation achieves the highest segmentation accuracy. This suggests that jointly applying spatial, intensity, and geometric perturbations provides stronger regularization, which is particularly beneficial under the noisy and low-contrast conditions of ultra-low-field MRI.

2.4   Training-Level Optimization: Patch Size and Batch Size

To adapt the model to the characteristics of ULF pediatric brain MRI, we conduct a series of controlled experiments to explore the effects of different input patch sizes and batch sizes on segmentation performance. Specifically, we evaluate two patch size configurations: $64 \times 64 \times 64$ and $128 \times 128 \times 128$, with corresponding batch sizes of 24 and 2, respectively, constrained by available GPU memory.

Empirical results show that using larger input patches significantly improves the model's ability to capture long-range spatial context and anatomical continuity, which are especially critical for segmenting low-contrast structures such as the basal ganglia. Among all configurations tested, the $128 \times 128 \times 128$ patch size with a batch size of 2 yields the most stable and accurate segmentation results across evaluation metrics.

Due to limited computational resources, we did not explore a wider range of patch and batch size combinations. Preliminary results indicate that larger patches might offer better segmentation performance, although further validation is required.

3   Experiments

3.1   Experimental Settings

Datasets. We conduct experiments on the dataset provided by the LISA 2025 Challenge (Task 2b), which focuses on the segmentation of the bilateral basal ganglia from ULF (0.064T) pediatric brain T2-weighted MRI scans. The dataset comprises 3D volumetric images acquired using the Hyperfine Swoop scanner, capturing early childhood brain anatomy with a reduced signal-to-noise ratio and resolution (Fig. 2).

Preprocessing. The LISA 2025 dataset was released after being co-registered to high-field MRI counterparts at an isotropic voxel size of $1.0 \times 1.0 \times 1.0$ mm^3. To ensure consistency and compatibility with our data-loading and patch extraction routines, we retained a uniform resampling step in our preprocessing pipeline. This step serves primarily as a verification of voxel grid alignment and metadata consistency across subjects, rather than a resolution change. All resampling operations used cubic B-spline interpolation for intensity images and nearest-neighbor interpolation for labels. Empirically, this additional verification step had negligible impact on image quality or segmentation accuracy, while ensuring stable preprocessing and reproducibility across runs. During both training and

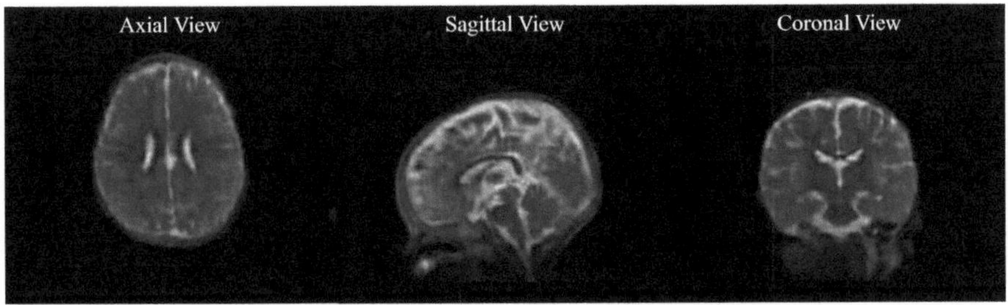

Fig. 2. A sample case from the LISA dataset.

inference, we randomly extract non-overlapping 3D patches of size $128 \times 128 \times 128$ from the preprocessed volumes.

Implementation Details. The model is implemented using PyTorch. During training, we use Stochastic Gradient Descent (SGD) as the optimizer with a learning rate of 0.01, momentum of 0.99, weight decay of 3×10^{-5}, and Nesterov acceleration enabled. A polynomial learning rate decay ("poly" scheduler) is employed to gradually reduce the learning rate over training epochs. The network is supervised using a standard cross-entropy loss function. The model is trained for 1000 epochs, with 250 iterations per epoch. The batch size is set to 2. All experiments are conducted on NVIDIA GTX 3090 GPUs with 24GB memory.

The proposed OMS-TCNet comprises approximately 59.5 million parameters, and the full-volume segmentation of a subject takes 24.17 s on average, including patch stitching and post-processing.

3.2   Results

Quantitative Results. We evaluate the segmentation performance of OMS-TCNet on the official validation set of the LISA 2025 Challenge Task 2b. The predicted results are submitted to the challenge website, and quantitative metrics are computed by the organizers. The evaluation includes Dice Similarity Coefficient (DSC), Hausdorff Distance (HD), 95th percentile Hausdorff Distance (HD95), Average Symmetric Surface Distance (ASSD), and Relative Volume Error (RVE) for each substructure of the basal ganglia, including the left/right caudate nucleus and lentiform nucleus.

As summarized in Table 1, OMS-TCNet achieves an average Dice score of 0.85 ± 0.05, demonstrating accurate and stable segmentation performance across all regions of interest. Notably, the method yields low HD95 and ASSD values, indicating precise boundary localization even under low-field imaging conditions.

Visualization Analysis. To qualitatively evaluate segmentation performance, we visualize the results on a representative subject randomly selected from our internally held-out validation set (9 : 1 split of the official training data). Because

Table 1. Quantitative segmentation results of OMS-TCNet on the LISA 2025 validation set. Metrics were computed separately for the left (L) and right (R) basal ganglia regions to capture potential asymmetry.

Structure	DSC	HD	HD95	ASSD	RVE
Caudate_L	0.84 ± 0.07	5.50 ± 6.20	1.62 ± 0.76	0.45 ± 0.22	0.11 ± 0.10
Caudate_R	0.85 ± 0.06	3.30 ± 1.29	1.51 ± 0.65	0.40 ± 0.20	0.11 ± 0.07
Lentiform_L	0.87 ± 0.06	2.97 ± 1.00	1.62 ± 0.76	0.51 ± 0.30	0.07 ± 0.04
Lentiform_R	0.86 ± 0.05	3.11 ± 1.64	1.79 ± 1.11	0.53 ± 0.22	0.09 ± 0.08
Average	**0.85 ± 0.05**	**3.72 ± 2.16**	**1.64 ± 0.72**	**0.47 ± 0.20**	**0.09 ± 0.06**

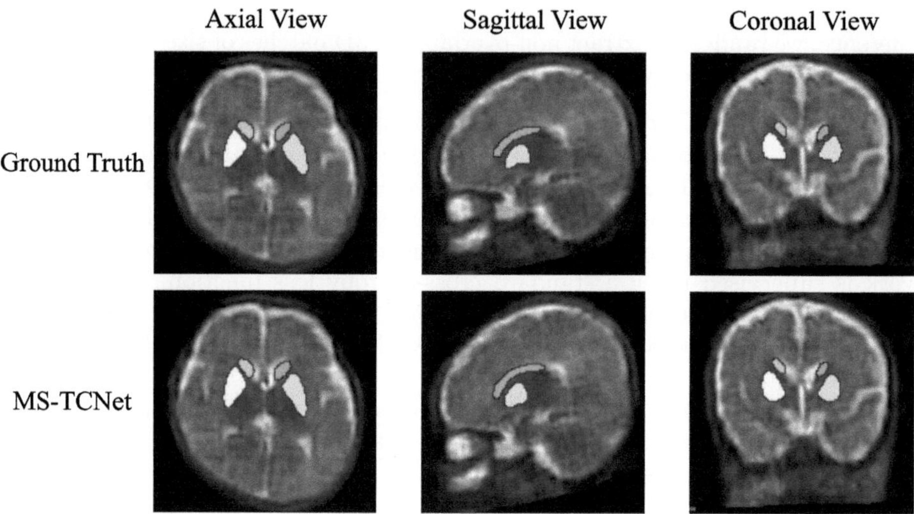

Fig. 3. Visual comparison between ground truth and OMS-TCNet predictions on a subject from the internal validation set.

the ground truth for the official LISA validation set is not publicly available, this internal validation subset allows for direct visual comparison.

As shown in Fig. 3, OMS-TCNet accurately captures the shape and boundaries of the basal ganglia, despite the low tissue contrast and image noise typical of 0.064T MRI scans. The predicted segmentation closely aligns with the ground truth, demonstrating the model's robustness in delineating deep brain structures in ULF settings.

4   Discussion

Our studies show that multi-scale feature modeling, coupled with appropriate data augmentation and training configurations, enables reliable segmentation under challenging imaging conditions. The effectiveness of full augmentation and the selection of an appropriate training configuration underscore the importance of comprehensive data variability and sufficient input coverage for achieving robust segmentation performance. Nonetheless, due to limited computational resources, only a small set of configurations was explored, and further validation on external datasets is warranted.

5   Conclusion

In this work, we explored the applicability of MS-TCNet to the task of segmenting the bilateral basal ganglia in 0.064T pediatric MRI. We adapted the original model to this challenging low-quality imaging context and introduced a series of practical improvements. These include a comprehensive data augmentation strategy based on the batchgenerators framework and empirical tuning of key training parameters such as patch size and batch size. Our experiments demonstrate that the optimized model, OMS-TCNet, produces accurate and reliable segmentation results on ULF brain scans, showing its potential for neuroimaging applications in resource-constrained clinical environments.

Acknowledgments. This work was supported in part by the National Natural Science Foundation of China under Grant 62201465.

Disclosure of Interests. The authors have no competing interests in the paper.

References

1. Aljabar, P., Heckemann, R.A., Hammers, A., Hajnal, J.V., Rueckert, D.: Multi-atlas based segmentation of brain images: atlas selection and its effect on accuracy. Neuroimage **46**(3), 726–738 (2009)
2. Ao, Y., Shi, W., Ji, B., Miao, Y., He, W., Jiang, Z.: Ms-tcnet: An effective transformer-CNN combined network using multi-scale feature learning for 3D medical image segmentation. Comput. Biol. Med. **170**, 108057 (2024)
3. Chen, G., et al.: Hybrid graph transformer for tissue microstructure estimation with undersampled diffusion mri data. In: International Conference on Medical Image Computing and Computer-Assisted Intervention, pp. 113–122. Springer (2022)
4. Chen, G., et al.: Em-trans: Edge-aware multimodal transformer for RGB-d salient object detection. IEEE Trans. Neural Netw. Learn. Syst. **36**(2), 3175–3188 (2024)
5. Deoni, S.C., et al.: Accessible pediatric neuroimaging using a low field strength MRI scanner. Neuroimage **238**, 118273 (2021)
6. Fischl, B., et al.: Whole brain segmentation: automated labeling of neuroanatomical structures in the human brain. Neuron **33**(3), 341–355 (2002)
7. Li, X., Wei, Y., Wang, L., Fu, S., Wang, C.: Msgse-net: Multi-scale guided squeeze-and-excitation network for subcortical brain structure segmentation. Neurocomputing **461**, 228–243 (2021)
8. Liu, Y., et al.: A low-cost and shielding-free ultra-low-field brain MRI scanner. Nat. Commun. **12**(1), 7238 (2021)

9. Mamah, D., Wang, L., Barch, D., de Erausquin, G.A., Gado, M., Csernansky, J.G.: Structural analysis of the basal ganglia in schizophrenia. Schizophr. Res. **89**(1–3), 59–71 (2007)
10. Patenaude, B., Smith, S.M., Kennedy, D.N., Jenkinson, M.: A Bayesian model of shape and appearance for subcortical brain segmentation. Neuroimage **56**(3), 907–922 (2011)
11. Ren, S., Zhou, D., He, S., Feng, J., Wang, X.: Shunted self-attention via multi-scale token aggregation. In: Proceedings of the IEEE/CVF Conference on Computer Vision and Pattern Recognition, pp. 10853–10862 (2022)
12. Ronneberger, O., Fischer, P., Brox, T.: U-net: Convolutional networks for biomedical image segmentation. In: International Conference on Medical Image Computing and Computer-assisted Intervention, pp. 234–241. Springer (2015)
13. Roy, A.G., Navab, N., Wachinger, C.: Concurrent spatial and channel 'squeeze & excitation'in fully convolutional networks. In: International Conference on Medical Image Computing and Computer-assisted Intervention, pp. 421–429. Springer (2018)
14. Sugino, T., Kin, T., Saito, N., Nakajima, Y.: Improved segmentation of basal ganglia from MR images using convolutional neural network with crossover-typed skip connection. Int. J. Comput. Assist. Radiol. Surg. **19**(3), 433–442 (2024)
15. Yang, J., Duncan, J.S.: 3D image segmentation of deformable objects with joint shape-intensity prior models using level sets. Med. Image Anal. **8**(3), 285–294 (2004)
16. Zhu, Y., Jiang, H., Cai, R., Chen, G.: Multi-label Mambaout for quality assessment of low-field pediatric brain mr images. In: MICCAI Challenge on Low Field Pediatric Brain Magnetic Resonance Image Segmentation and Quality Assurance, pp. 3–11. Springer Nature Switzerland Cham (2024)
17. Zhu, Y., Lv, H., Chen, G., Zhang, Z., Jiang, H., Xia, Y.: Hybrid graph mamba: Unlocking non-euclidean potential for accurate polyp segmentation. In: International Conference on Medical Image Computing and Computer-Assisted Intervention, pp. 277–286. Springer (2025)

Tasks 2a and 2b - Automatic Hippocampal and Basal Ganglia Segmentation form Ultra-Low Field MRI

Segmenting Brain Regions in Low Field Pediatric Brain MR Images Using (Symmetric) NnU-Net ResEnc

Jan Nikolas Morshuis[1(✉)], Matthias Hein[1], and Christian F. Baumgartner[1,2]

[1] University of Tübingen, Tübingen, Germany
`nikolas.morshuis@uni-tuebingen.de`
[2] University of Lucerne, Lucerne, Switzerland

Abstract. The segmentation of low-field pediatric brain MR images is an important topic, as it can show the development of the pediatric brain. At the same time, the low cost and maintenance required by the low-field scanners make the technology more accessible in wider parts of the world compared to high-field MRI-scanners. The wider accessibility allows to understand the pediatric brain development in different regions of the world, thereby allowing to better understand the effects of, for example, nutrition for the brain development. Despite these advantages, automatically segmenting low-field MRI scans can be challenging: The internal brain structure can be hard to recognize in the low-field MRI images, the ground-truth segmentation can be unprecise due to registration errors between high-field and low-field MRI images, and the segmentation can potentially be different for the hippocampus region depending on whether it is the left or right hippocampus. In order to still be able to achieve the best possible segmentation scores, we try to increase the probability that our predicted segmentation is close to the ground-truth segmentation. To achieve this, we focus on extending the data-augmentation and make use of an ensemble of networks, of which we take the average as our final prediction. Doing so led to high-scores for the task 2 of the LISA challenge. The code is available at https://github.com/NikolasMorshuis/nnUNet-LISA-Challenge.git.

Keywords: Segmentation · Low Field MRI · Brain imaging

1 Introduction

Magnetic Resonance Imaging (MRI) plays a crucial role in the non-invasive assessment of brain structure and development, particularly in pediatric populations. High-resolution brain MRI is essential for diagnosing developmental abnormalities, planning surgical interventions, and studying neurodevelopmental trajectories. However, access to high-field MRI systems (e.g., 3T or higher) remains limited in many low-resource settings due to their cost, infrastructure requirements, and limited portability. As a result, low-field MRI scanners—which

© The Author(s) 2026
N. Lepore and M. G. Linguraru (Eds.): LISA 2025, LNCS 16411, pp. 41–49, 2026.
https://doi.org/10.1007/978-3-032-14417-1_4

typically operate at 0.55T or below—have garnered growing interest as a more affordable and accessible alternative [9,12]

Despite their practical advantages, low-field MRI scans present significant technical challenges. These include reduced signal-to-noise ratio (SNR), lower spatial resolution, and increased susceptibility to artifacts. These limitations particularly impact automated image analysis tasks such as brain tissue segmentation, which require high image quality and robust contrast between tissue classes. Pediatric imaging compounds these difficulties: the developing brain exhibits rapidly evolving anatomical structures, age-dependent tissue properties, and greater inter-subject variability compared to adult brains [5]. These factors demand segmentation algorithms that are both noise-tolerant and adaptable across different stages of brain development.

Accurate segmentation of pediatric brain MRI scans is essential for a variety of clinical and research applications. In clinical settings, segmentation maps support volumetric assessments used to diagnose and monitor developmental disorders, such as hydrocephalus or cerebral atrophy [4]. In research contexts, automated segmentation enables population-level studies of brain growth, allowing for longitudinal analysis of brain morphology in relation to cognitive and behavioral outcomes [8].

The LISA 2025 challenge addresses these issues by providing a curated dataset of low-field pediatric brain MRI scans, along with a standardized evaluation framework. The goal is to benchmark segmentation algorithms under realistic imaging constraints. In this work, we present a deep learning-based method tailored to the characteristics of low-field pediatric MRI. Our method is based on the nnU-Net framework [6]. We adjusted the data-annotation given by the challenge to allow for more data-augmentation, and achieve top performance on the validation leaderboard.

2   Related Work

Automated brain segmentation has been a long-standing problem in medical image analysis, with early methods relying on atlas-based registration, tissue modeling, and intensity thresholding [1,3]. In recent years, convolutional neural networks (CNNs) have emerged as the dominant approach, led by the introduction of the U-Net architecture [11], which provides strong localization performance while capturing multiscale context. Many variants of U-Net have been developed to improve robustness to noise, preserve boundary details, or handle domain shifts [6,15].

Low-field MRI has recently gained attention due to its potential to improve imaging access in underserved settings. However, segmentation performance on low-field scans remains an underexplored area. Most existing segmentation benchmarks and models are trained on high-field (e.g., 3T) data and struggle to generalize to lower-field images with reduced resolution and contrast [14]. Recent work has explored domain adaptation and harmonization techniques to address this, including intensity standardization, style transfer, and adversarial training [2].

In another challenge called the K2S-challenge [13] presented at MICCAI 2022, participants also had to segment degraded MRI data. The goal was to provide segmentations for 8x-undersampled MRI data. Interestingly, the best two methods - although conceptually very different - perform similarly well: For each of the six segmentation categories, the difference in Dice score between the methods was ≤ 0.01. Furthermore, Morshuis et al. [10] have shown that most state-of-the-art models for the segmentation of undersampled MRI data perform very comparably. The finding suggests that there is potentially some kind of upper limit in segmentation accuracy when dealing with degraded MRI images and imperfect human segmentations as ground-truth. We therefore expect the results of the LISA 2025 challenge to also be rather similar in terms of Dice score.

A recent paper by Isensee et al. [7] has shown that despite the fact that currently there are many new methods released that claim to improve on some segmentation tasks, methods based on the nnU-Net framework usually outperform these new methods when common validation shortcomings are removed. The results of the paper indicate, that the field of medical image segmentation has an ongoing innovation bias towards newer and potentially more complex methods, even though these methods often do not perform better than the previous methods that are often times also simpler and usable for a wider range of segmentation problems. The authors have also shown that one of their variants of the original nnU-Net called nnU-Net ResEnc often achieves higher scores compared to the standard nnU-Net method. In this paper, we will therefore also make use of the nnU-Net ResEnc implementation for task 2 and find that it can indeed provide good segmentations, achieving scores that are among the highest for one of the tasks.

3   Method

Our method is based on the nnU-Net [6] framework, as this framework has proven to work well for a large amount of different medical segmentation challenges. More specifically, we decide to make use of the nnU-Net ResEnc method, as it has been shown [7] that this method often outperforms the standard nnU-Net. We chose the Large (L) variant for the symmetric nnU-Net, as it often provides better scores than the M and XL variants on small to medium large datasets.

The baseline nnU-Net ResEnc was originally intended as a simple method to indicate which region is where, in order to map the results of the symmetric nnU-Net ResEnc to the correct labels to make the prediction comparable to the ground-truth. We therefore only used the M variant for this method. With this method, the mirroring augmentation needs to be deactivated, as otherwise the left and right labels can be difficult to distinguish for the model.

Both the symmetric nnU-Net ResEnc and the baseline nnU-Net ResEnc were trained using 5-fold cross validation. During inference, the predicted segmentation probabilities were averaged for each method individually in order to obtain the prediction from the ensemble.

As the boundaries of the ground-truth segmentation are hardly visible - if at all- for the untrained eye, we think it is important that the patch-size of the inputs during training is relatively large. This assures that the model can learn to understand the context of the image and knows in which regions the hippocampi and basal ganglia are usually located. The large patch-sizes therefore also prevent the model to be unable to locate the given patch-region to the right segmentation class, as potentially the information within the patch is not sufficient to assign the correct segmentation class with enough confidence.

Another adjustment that we have made is to change the ground-truth segmentation class annotations slightly: As the brain is generally relatively symmetric, we have adjusted the segmentation classes such that e.g. the left and right hippocampus have the same class. This way the method only needs to learn 3 classes that are relevant for the evaluation: 1. Hippocampus, 2. Caudate Nucleus, 3. Lentiform Nucleus. The extra distinction of classes between left and right is therefore not done during the training process. The prediction is only later post-processed to distinguish between left and right class, see below. Our method allows the model to learn the concept of the different classes, without the necessity to distinguish between left and right. It also allows to use more data-augmentation like mirroring in all directions. This extra data-augmentation is important when dealing with medium amounts of data like the 79 MRI scans provided by the LISA 2025 challenge.

As our main framework only predicts the three classes (Hippocampus, Caudate Nucleus, Lentiform Nucleus), yet the evaluation requires a distinction between the left and right segmentations, we also trained another segmentation network that is able to distinguish between the left and the right segmentation class. We make use of this second network to determine whether an e.g. Caudate Nucleus voxel should belong to either the left or right Caudate Nucleus. To do this we calculate the distance of every predicted voxel of prediction A to their nearest voxel in prediction B belonging to either the left or the right part of the corresponding class.

For the task 2a hippocampus segmentation, we found that there is a significant difference between the left and the right hippocampus segmentation in terms of Dice score. This potentially suggests that there might indeed be some difference between the segmentations of the left and right hippocampus. We therefore do not utilize our symmetric network for these regions. We instead train an extra network that is only training using the labels for left and right hippocampus for the task 2a, thereby neglecting the remaining labels.

4   Results

We found our symmetric approach to work quite well on the task 2b of segmenting the bilateral basal ganglia. As all available data has been used to train the models, we can only report the scores using the validation set provided by the challenge organizers and the evaluation results provided by the available table. We can test our symmtric method against the 'vanilla' method of just training a UNet (Fig. 2).

As shown in Table 1, the difference between both methods is small. Both methods have very similar Dice scores (0.87) and also similar ASSD scores (0.42).

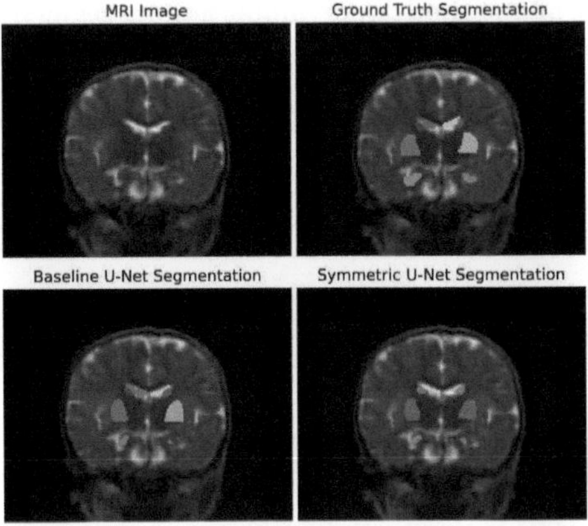

Fig. 1. Example of Ground truth segmentation vs. our Baseline U-Net vs. our Symmetric U-Net. The Symmetric U-Net predicts the same label for the left and the right side. Note that the symmetric U-Net does not predict the ventricles, as these were not part of the challenge tasks.

Table 1. Comparison of the models for basal ganglia segmentation (Task 2b) using Dice coefficient, Hausdorff Distance (HD), and 95th percentile Hausdorff Distance (HD95), Average Symmetric Surface Distance (ASSD) and Relative Volume Error (RVE). Higher Dice and lower HD/HD95/ASSD/RVE indicate better performance.

Model	Dice	HD	HD95	ASSD	RVE
Sym. U-Net	**0.87±0.05**	**2.76±0.58**	**1.46±0.49**	**0.42±0.19**	0.08±0.03
Base U-Net	**0.87±0.05**	2.94±0.91	1.52±0.68	**0.42±0.21**	**0.06±0.03**

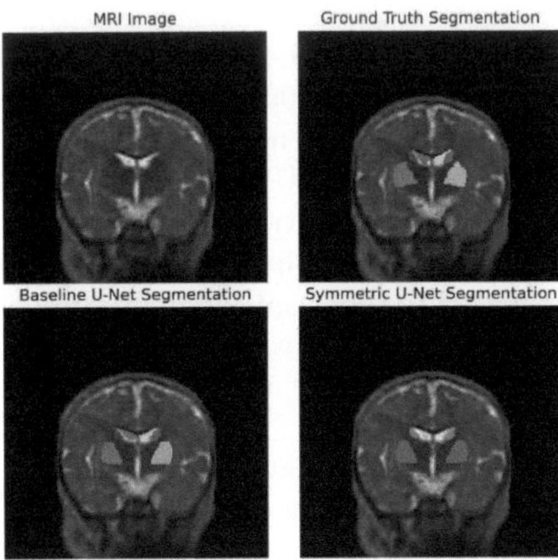

Fig. 2. Another example of Ground Truth vs. Baseline vs. Symmetric U-Net Segmentations.

Table 2. Comparison of the models for hippocampus segmentation (Task 2a) using Dice coefficient, Hausdorff Distance (HD), and 95th percentile Hausdorff Distance (HD95), Average Symmetric Surface Distance (ASSD) and Relative Volume Error (RVE). Higher Dice and lower HD/HD95/ASSD/RVE indicate better performance.

Model	Dice	HD	HD95	ASSD	RVE
Sym. U-Net	0.71±0.19	5.56±7.89	1.96±1.49	0.74±0.90	0.15±0.12
Base U-Net	**0.72±0.19**	**5.43±7.80**	1.91±1.49	**0.69±0.83**	**0.14±0.09**
Hip. U-Net	**0.72±0.19**	5.49±8.01	**1.89±1.45**	0.71±0.90	0.16±0.09

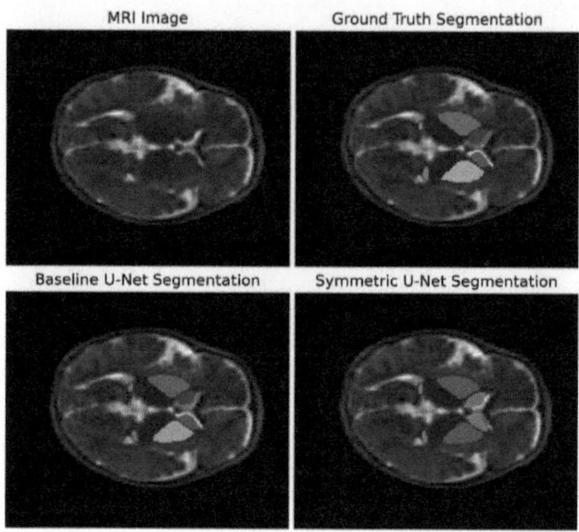

Fig. 3. Example ground-truth segmentation and predictions from our baseline and symmetric U-Net models. Example is shown from the axial direction.

The symmetric nnUNet is better on both HD and HD95, while the baseline nnUNet achieves better RVE scores (0.06 vs. 0.08). As of now it is unclear if the additional complexity of the Symmetric nnUNet that requires multiple additional steps in comparison to the baseline nnUNet is worth it, as each additional step is a potential source of error in production.

For the task 2a of segmenting the hippocampus, however, the segmentations achieve lower scores compared to the basal ganglia results of the task 2b discussed above. For that reason, we also train a 'specialized' model that only sees the hippocampus data during training that is termed "Hip. U-Net". The results are shown in Table 2.

It can be seen that the results are relatively close together for all three models, with e.g. the Dice score having deviating only by ±0.01. However, our specialized model Hip. U-Net that has only seen hippocampus segmentations during training does not perform much better compared to the other more general variants. Because Base U-Net achieves the best HD, ASSD and RVE scores, we will select the Base U-Net for the task 2a during the challenge. Example segmentations of the hippocampus region can be seen in Fig. 4.

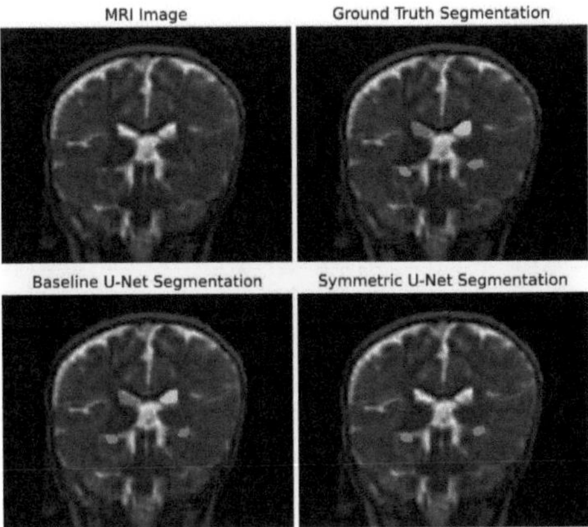

Fig. 4. Example of the hippocampus segmentation. Our predictions tend to be slightly rounder and more oval-shaped, while in the ground truth the shape tends to be more complex. Again, note that the ventricles are not predicted by Symmetric U-Net, as these are not relevant for the challenge tasks.

4.1 Data Properties

As can be seen in Figs. 1 and 5, the Lentiform Nucleus is sometimes segmented with an abrupt end, where the segmentation seems to end on a single slice. This behaviour is partly copied by the learned segmentation models, as can also be seen in Fig. 3. At the same time, the ground-truth label in Fig. 3 appears to have a more natural shape compared to the ground-truth label in Fig. 3. As for many segmentation tasks, the segmentations are often not-optimal and the limited time often available from doctors can lead to shortcuts in labeling, that

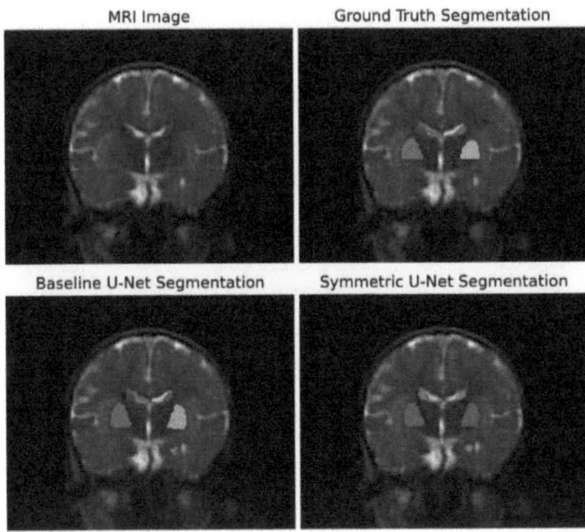

Fig. 5. Example of Lentiform Nucleus segmentation ending on a single slice. Our predicted segmentations are slightly rounder.

gets also learned by the models themselves in order to reach the best possible scores given the imperfect data.

5   Discussion

We have shown that the adaption of a fairly standard method for semantic segmentation can achieve relatively high scores. As of the time of writing, the Sym U-Net as well as the Base U-Net achieve the two highest Dice scores in the leaderboard for the task 2b, despite mainly being an off-the-shelf solution for medical segmentation problems.

However, we were not able to find a model for the task 2a of segmenting the hippocampus regions, that achieves the highest Dice scores in the competition. A potential reason that our Sym. U-Net did not achieve the highest scores from the models that we tested could be, that the scores for left and right hippocampus are indeed different for most models in Table 2. Our best performing model on this task, the Base U-Net, achieves an average Dice score of 0.72, yet the Dice score for the left hippocampus is only 0.69 while the right hippocampus achieves Dice scores of 0.76. This indicates that there are systematic differences between the left and right hippocampus annotations that are difficult to learn when using the symmetric U-Net model.

The difference in labels as shown in Figs. 1 and 3 can lead to a large difference in the final score. Our model seems to learn that especially the Lentiform Nucleus can end on a single axial slice and is therefore mimicking the behavior it has seen during training. This is usually a behavior that we like to see, as we commonly assume the training data to be the ground-truth and the gold-standard in machine learning tasks. However, if during testing the predictions are compared to perfect segmentations that were annotated with more time and care, the scores of a model that has learned to mimic these imperfections might drop. However, if the test-set is similar to the training set which is expected, we believe our model to perform relatively well.

Acknowledgments. Funded by the Deutsche Forschungsgemeinschaft (DFG, German Research Foundation) under Germany's Excellence Strategy - EXC number 2064/1 - Project number 390727645. The authors thank the International Max Planck Research School for Intelligent Systems (IMPRS-IS) for supporting JNM.

Disclosure of Interests. The authors have no competing interests to declare.

References

1. Ashburner, J., Friston, K.J.: Unified segmentation. Neuroimage **26**(3), 839–851 (2005)
2. Chartsias, A., Joyce, T., Dharmakumar, R., Tsaftaris, S.A.: Adversarial image synthesis for unpaired multi-modal cardiac data. In: International Workshop on Simulation and Synthesis in Medical Imaging, pp. 3–13. Springer (2017)

3. Fischl, B.: Freesurfer. Neuroimage **62**(2), 774–781 (2012)
4. Fischl, B., et al.: Whole brain segmentation: automated labeling of neuroanatomical structures in the human brain. Neuron **33**(3), 341–355 (2002)
5. Ghosh, S.S., et al.: Evaluating the validity of volume-based and surface-based brain image registration for developmental cognitive neuroscience studies in children 4 to 11 years of age. Neuroimage **53**(1), 85–93 (2010)
6. Isensee, F., Jaeger, P.F., Kohl, S.A., Petersen, J., Maier-Hein, K.H.: nnu-net: a self-configuring method for deep learning-based biomedical image segmentation. Nat. Methods **18**(2), 203–211 (2021)
7. Isensee, F., et al.: nnu-net revisited: A call for rigorous validation in 3d medical image segmentation. In: International Conference on Medical Image Computing and Computer-Assisted Intervention, pp. 488–498. Springer (2024)
8. Knickmeyer, R.C., et al.: A structural MRI study of human brain development from birth to 2 years. J. Neurosci. **28**(47), 12176–12182 (2008)
9. Marques, J.P., Simonis, F.F., Webb, A.G.: Low-field MRI: an MR physics perspective. J. Magn. Reson. Imaging **49**(6), 1528–1542 (2019)
10. Morshuis, J.N., Hein, M., Baumgartner, C.F.: Understanding benefits and pitfalls of current methods for the segmentation of undersampled MRI data (2025)
11. Ronneberger, O., Fischer, P., Brox, T.: U-net: Convolutional networks for biomedical image segmentation. In: International Conference on Medical Image Computing and Computer-assisted Intervention, pp. 234–241. Springer (2015)
12. Sarracanie, M., Salameh, N.: Low-field MRI: how low can we go? A fresh view on an old debate. Front. Phy. **8**, 172 (2020)
13. Tolpadi, A.A., et al.: K2s challenge: from undersampled k-space to automatic segmentation. Bioengineering **10**(2), 267 (2023)
14. Wang, G., et al.: Deepigeos: a deep interactive geodesic framework for medical image segmentation. IEEE Trans. Pattern Anal. Mach. Intell. **41**(7), 1559–1572 (2018)
15. Zhou, Z., Rahman Siddiquee, M.M., Tajbakhsh, N., Liang, J.: Unet++: A nested u-net architecture for medical image segmentation. In: International Workshop on Deep Learning in Medical Image Analysis, pp. 3–11. Springer (2018)

Segmenting Infant Brains Across Magnetic Fields: Domain Randomization and Annotation Curation in Ultra-low Field MRI

Vladyslav Zalevskyi[1,2]($\boxtimes$), Dondu-Busra Bulut[1,2], Thomas Sanchez[1,2], and Meritxell Bach Cuadra[1,2]

[1] Department of Radiology, Lausanne University Hospital and University of Lausanne (UNIL), Lausanne, Switzerland
[2] CIBM Center for Biomedical Imaging, Lausanne, Switzerland
`vladyslav.zalevskyi@unil.ch`

Abstract. Early identification of neurodevelopmental disorders relies on accurate segmentation of brain structures in infancy, a task complicated by rapid brain growth, poor tissue contrast, and motion artifacts in pediatric MRI. These challenges are further exacerbated in ultra-low-field (ULF, 0.064 T) MRI, which, despite its lower image quality, offers an affordable, portable, and sedation-free alternative for use in low-resource settings. In this work, we propose a domain randomization (DR) framework to bridge the domain gap between high-field (HF) and ULF MRI in the context of the hippocampi and basal ganglia segmentation in the LISA challenge. We show that pre-training on whole-brain HF segmentations using DR significantly improves generalization to ULF data, and that careful curation of training labels, by removing misregistered HF-to-ULF annotations from training, further boosts performance. By fusing the predictions of several models through majority voting, we are able to achieve competitive performance. Our results demonstrate that combining robust augmentation with annotation quality control can enable accurate segmentation in ULF data. Our code is available at https://github.com/Medical-Image-Analysis-Laboratory/lisasegm

Keywords: Ultra-low-field MRI · Pediatric imaging · Domain Randomization · Quality control

1 Introduction

Accurate segmentation of brain structures in early childhood is essential for studying typical neurodevelopment and identifying early signs of neurological

---

T. Sanchez and M. Bach Cuadra—Shared senior authorship.

Supplementary Information The online version contains supplementary material available at https://doi.org/10.1007/978-3-032-14417-1_5.

N. Lepore and M. G. Linguraru (Eds.): LISA 2025, LNCS 16411, pp. 50–62, 2026.
https://doi.org/10.1007/978-3-032-14417-1_5

disorders. Deep gray matter regions such as the basal ganglia and hippocampi are particularly important due to their roles in motor control, cognition, and memory—functions commonly impacted in conditions like ADHD and autism spectrum disorders [1–3]. However, existing deep learning segmentation models, largely trained on adult high-field (HF) MRI, often perform poorly on pediatric low-field (LF, 0.1–1 T) and ultra-low-field (ULF, <0.1T) scans. This generalization gap stems from anatomical differences between adults and infants, field-strength-dependent contrast changes, and frequent motion artifacts in pediatric populations [4]. These challenges are amplified in low-resource settings, where access to HF scanners (1.5 T/3 T) is limited [5,6]. ULF MRI systems, such as the 0.064 T Hyperfine scanner, offer a portable and cost-effective alternative for pediatric imaging [7–10]. Yet, their lower signal-to-noise ratio and reduced spatial resolution demand methods that are explicitly adapted to this imaging regime.

Domain randomization (DR) has emerged as a promising strategy to improve generalization across MRI modalities, contrasts, and populations. Methods such as `SynthSeg` [11,12] augment label-derived images with randomized intensity and spatial transformations, enabling robust performance even on highly heterogeneous or low-quality data [13]. In particular, DR has recently been applied to fetal and neonatal imaging [14,15], but to our knowledge, no prior work has addressed the joint challenge of pediatric anatomy and ULF domain shift.

In this paper, we investigate DR-based pre-training for segmenting deep gray matter structures in infant ULF MRI, as part of the LISA challenge. We evaluate how DR can support cross-domain generalization from HF data and examine the impact of annotation quality on fine-tuning performance. Specifically, we show that: **(1)** Domain randomization enables HF-to-ULF annotation transfer, **(2)** Pre-training on whole-brain annotations boosts task-specific segmentation, **(3)** Filtering out misaligned labels improves robustness, and **(4)** Model ensembling further increases the performance of our models.

We believe that these observations will inform the design of segmentation pipelines that are both accurate and practical, particularly in low-resource clinical environments where ULF MRI offers a promising solution for early neurodevelopmental assessment.

2   Methods

2.1   Datasets

We used three datasets with neonatal and infant brain images and segmentations in this work. Two publicly available HF datasets were used for pre-training (dHCP [16] and BOBs [17]), and the LISA challenge dataset [4] is used for fine-tuning. A summary is provided in Table 1.

Baby Open Benchmark Segmentations (BOBs). The BOBs dataset [18] was acquired on a high-field 3T Siemens Prisma system at the University of Minnesota, United States, following a Baby Connectome Project (BCP) protocol [19]. This subset includes only healthy subjects for which extensive, manually curated segmentations are available. The segmentations are aligned to both

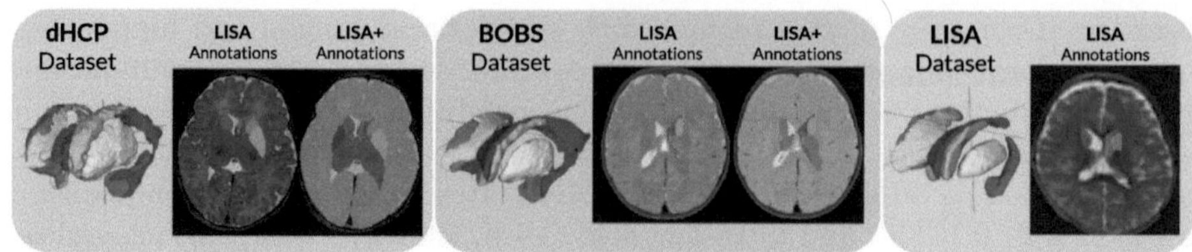

Fig. 1. Overview of the datasets and annotation schemes used in this study. Some variations are found across dataset annotation definition: e.g. in both dHCP and BOBs the lateral ventricle label extends further down toward the hippocampi).

Table 1. Summary of the datasets used for pre-training (dHCP and BOBs) and fine-tuning (LISA challenge dataset).

Dataset	Scanner	Age range	Conditions	Size	Resolution
BOBs [17]	3T Siemens Prisma	1–9 months	Healthy	51 infants, 71 sessions	0.8 mm^3
dHCP [16]	3T Philips Achieva	23–44 gestational weeks	Healthy	783 infants, 887 sessions	0.5 mm^3
LISA [4]	0.064T Hyperfine SWOOP	4.5 weeks–16 months	Healthy	79 (training), 12 (validation)	1 mm^3

T1-weighted and T2-weighted scans that were acquired and include labels for cerebral gray matter, white matter, and 23 subcortical structures.

Developing Human Connectome Project (dHCP). The dHCP dataset [20] was acquired at the Evelina Newborn Imaging Centre, King's College London, using a 3T Philips Achieva scanner optimized for neonatal imaging. We used the T2-weighted images with the provided automated segmentations into 9 tissue classes and 87 regions [21]. Note that while the data is rich, the dHCP dataset is centered around the perinatal period, compared to the LISA challenge data covering subjects up to 16 months.

LISA Dataset. The Low-field Pediatric Brain Magnetic Resonance Image Segmentation and Quality Assurance (LISA) dataset [4] was collected from healthy neonates at three sites: the University of Cape Town (South Africa), Makerere University (Uganda), and Aga Khan University Hospital (Pakistan). Imaging was performed using portable ultra-low-field (ULF) 0.064 T Hyperfine SWOOP scanners. For this work, we tackled the second task of the LISA challenge, which consisted in segmenting super-resolution-reconstructed [22] T2-weighted images across 8 structures: left and right hippocampi, lateral ventricles, caudate nuclei, and lentiform nuclei. Two sets of ground truth labels were provided: initial labels done on the ULF images (GT_{LF}), and labels created on the high-field (1.5T or 3T) image counterparts and subsequently propagated to the ULF images through nine-point linear co-registration (GT_{HF}). The second set of labels was the one used as ground truth in the challenge.

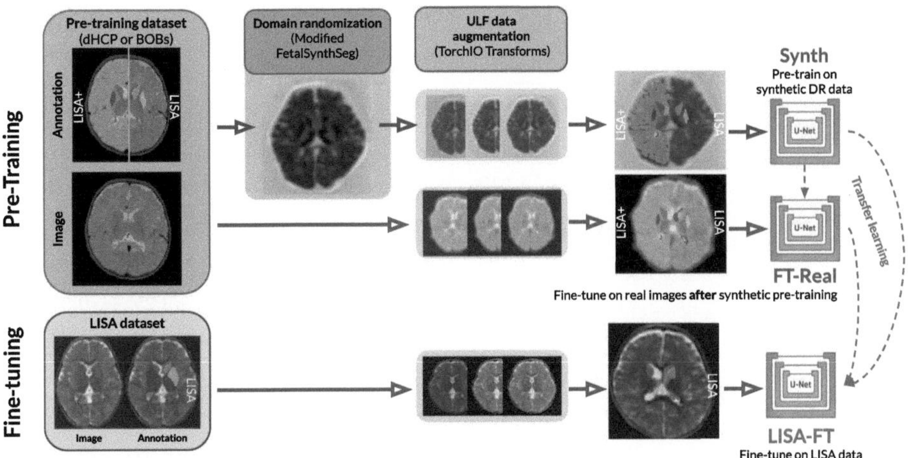

Fig. 2. Overview of training modes used in this study. We start with a pre-training step yielding a `Synth` model that can then be fine-tuned on the real images from the pre-training data to yield `FT-Real`. These models are then fine-tuned on LISA data for evaluation. Red arrows show the use of synthetic data, blue arrows show real images, and green arrows show fine-tuning paths.

2.2   Data Harmonization and Pre-Processing

We first resampled every dataset's images to an isotropic voxel size of 1 mm^3, matching the resolution of the LISA dataset. Because the pre-training datasets (dHCP and BOBs) contained a much denser set of annotation labels than those defined in LISA, we pre-trained our models using two annotation variants, as illustrated in Fig. 1:

1. **LISA:** We selected from the pre-training datasets only the labels present in the LISA dataset. Both datasets contain the required classes for the LISA challenge, although with slightly different definitions.
2. **LISA+:** We used the dense structural information from dHCP and BOBs to pre-train our models. As dHCP and BOBs contained many labels, we grouped their annotations in 6 groups covering the entire brain: white matter (WM), cortical gray matter (GM), cerebrospinal fluid (CSF – dHCP only), cerebellum, brainstem, and deep GM (excluding the basal ganglia and hippocampi). These labels were used in addition to the 8 LISA labels during training.

2.3   Domain Randomization

In this work, we used domain randomization as the core of our training procedure. DR relies on a generating procedure producing randomized intensity images with paired annotations that are then used to train a segmentation model.

Table 2. Pre-training results on LISA validation dataset. Best value in each training group is in **bold** (`NormAvg.` only).

Task	Pretrain.	Annotation	Training	NormAvg ↓	DSC ↑	HD ↓	HD95 ↓	ASSD ↓	RVE ↓
2A–HIPPOCAMPUS	BOBs	LISA	FT-Real	**1.385**	$0.56_{\pm 0.16}$	$12.46_{\pm 7.92}$	$6.89_{\pm 1.34}$	$1.52_{\pm 0.88}$	$0.28_{\pm 0.18}$
			Synth	1.610	$0.54_{\pm 0.14}$	$10.96_{\pm 6.97}$	$5.92_{\pm 1.55}$	$1.61_{\pm 0.75}$	$0.60_{\pm 0.25}$
		LISA$_+$	FT-Real	**1.240**	$0.57_{\pm 0.14}$	$9.97_{\pm 7.35}$	$4.89_{\pm 1.30}$	$1.40_{\pm 0.76}$	$0.45_{\pm 0.22}$
			Synth	1.809	$0.52_{\pm 0.13}$	$11.16_{\pm 7.19}$	$5.94_{\pm 1.47}$	$1.67_{\pm 0.62}$	$0.74_{\pm 0.32}$
	dHCP	LISA	FT-Real	**3.532**	$0.32_{\pm 0.08}$	$18.34_{\pm 6.98}$	$12.49_{\pm 1.69}$	$3.25_{\pm 0.55}$	$0.91_{\pm 0.42}$
			Synth	4.169	$0.25_{\pm 0.06}$	$19.87_{\pm 6.65}$	$14.18_{\pm 2.19}$	$3.98_{\pm 0.67}$	$1.06_{\pm 0.42}$
		LISA$_+$	FT-Real	2.881	$0.38_{\pm 0.09}$	$18.16_{\pm 7.52}$	$12.35_{\pm 2.52}$	$2.93_{\pm 0.84}$	$0.45_{\pm 0.28}$
			Synth	**2.841**	$0.40_{\pm 0.09}$	$16.75_{\pm 6.87}$	$10.89_{\pm 1.56}$	$2.71_{\pm 0.53}$	$0.68_{\pm 0.32}$
2B–BASAL GANGLIA	BOBs	LISA	FT-Real	**1.011**	$0.76_{\pm 0.04}$	$5.83_{\pm 1.11}$	$3.52_{\pm 1.08}$	$0.92_{\pm 0.23}$	$0.16_{\pm 0.06}$
			Synth	1.088	$0.74_{\pm 0.05}$	$6.08_{\pm 1.06}$	$3.65_{\pm 0.98}$	$0.99_{\pm 0.23}$	$0.14_{\pm 0.10}$
		LISA$_+$	FT-Real	1.469	$0.71_{\pm 0.05}$	$6.68_{\pm 1.56}$	$4.50_{\pm 1.20}$	$1.13_{\pm 0.26}$	$0.22_{\pm 0.06}$
			Synth	**1.460**	$0.73_{\pm 0.05}$	$7.02_{\pm 1.02}$	$4.52_{\pm 1.04}$	$1.11_{\pm 0.21}$	$0.24_{\pm 0.10}$
	dHCP	LISA	FT-Real	**3.669**	$0.53_{\pm 0.04}$	$10.91_{\pm 1.06}$	$8.03_{\pm 1.09}$	$2.20_{\pm 0.28}$	$0.68_{\pm 0.20}$
			Synth	4.070	$0.50_{\pm 0.05}$	$11.83_{\pm 1.37}$	$9.10_{\pm 1.15}$	$2.45_{\pm 0.27}$	$0.72_{\pm 0.20}$
		LISA$_+$	FT-Real	**3.915**	$0.57_{\pm 0.05}$	$11.66_{\pm 4.01}$	$8.53_{\pm 4.04}$	$3.01_{\pm 4.00}$	$0.66_{\pm 0.16}$
			Synth	4.274	$0.51_{\pm 0.08}$	$12.37_{\pm 3.67}$	$9.66_{\pm 3.85}$	$3.23_{\pm 3.36}$	$0.63_{\pm 0.20}$

Data Generation. For the generation procedure, we used the synthetic data generator from the `FetalSynthSeg` framework by Zalevskyi *et al.* [15] and adapted it to the ULF setting. This publicly available generator[1] was tailored for fetal populations using T2w images and was easy to adapt to the pediatric population of the LISA challenge.

Most parameters of the generator were kept unchanged: rigid and non-rigid spatial deformations, intensity and contrast randomization, as well as intensity generation were kept identical to previous work on low-field (0.55 T) fetal brain segmentation [15] as we found these parameters to readily apply to the ULF domain. We changed the resolution resampling step to downsample the generated volume to an anisotropic volume with slice thickness between 1 and 5 mm, before interpolating it back to an isotropic resolution of 1 mm^3 (simulating reconstruction of ULF images). To account for ULF-specific artifacts not simulated by `FetalSynthSeg`, we additionally applied a set of randomized augmentations that simulate common artifacts in ULF MRI: random k-space motion artifacts, k-space ghosting, and spiking artifacts [23]. These artifacts improved the robustness of our model—even though it remains vulnerable to artifacts not simulated during training—while the domain randomized approach allowed it to perform well even on cases with very low tissue contrast.

---

[1] https://github.com/Medical-Image-Analysis-Laboratory/fetalsyngen.

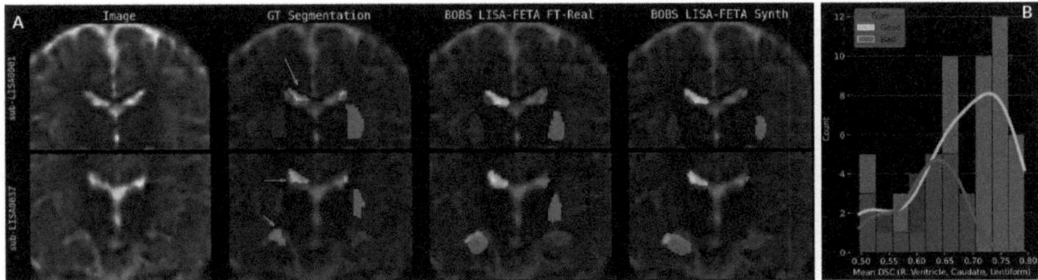

Fig. 3. **(A)** Examples of annotation drift caused by registration errors in the ground-truth (GT) segmentations of two representative LISA cases, compared with the outputs from our pre-trained models. **(B)** Distribution of the mean DSC for several critically affected right-sided labels across the training dataset, comparing the LISA GT annotations with the BOBs +LISA$_+$+**FT-Real** model. Note the bimodal distribution, reflecting the difference in Dice scores between well-registered and poorly registered images. The images shown in panel (A) have mean DSC values of 0.53 and 0.57.

Table 3. Quantitative metrics by task, annotation, and training type (training on exclusively images that we deemed **good** or **all** training data). All models trained are on BOBs dataset.

Task	Annot.	Training	Type	NormAvg ↓	DSC ↑	HD ↓	HD95 ↓	ASSD ↓	RVE ↓
2A–Hippocampus	LISA	FT-Real	All	0.268	$0.67_{\pm0.18}$	$6.41_{\pm8.21}$	$2.45_{\pm1.24}$	$0.86_{\pm0.81}$	$0.17_{\pm0.08}$
	LISA	FT-Real	Good	**0.215**	$0.68_{\pm0.19}$	$6.16_{\pm8.05}$	$2.25_{\pm1.67}$	$0.87_{\pm1.08}$	$0.16_{\pm0.07}$
	LISA	Synth	All	**0.183**	$0.70_{\pm0.19}$	$6.06_{\pm7.74}$	$2.23_{\pm1.56}$	$0.81_{\pm0.95}$	$0.18_{\pm0.07}$
	LISA	Synth	Good	0.231	$0.69_{\pm0.18}$	$6.27_{\pm8.06}$	$2.31_{\pm1.64}$	$0.85_{\pm1.05}$	$0.19_{\pm0.07}$
	LISA$_+$	FT-Real	All	0.199	$0.68_{\pm0.18}$	$5.83_{\pm7.88}$	$2.15_{\pm1.28}$	$0.80_{\pm0.77}$	$0.17_{\pm0.07}$
	LISA$_+$	FT-Real	Good	**0.118**	$0.69_{\pm0.18}$	$5.73_{\pm7.79}$	$2.02_{\pm1.42}$	$0.80_{\pm0.90}$	$0.12_{\pm0.08}$
2B–BaGa	LISA	FT-Real	All	0.150	$0.85_{\pm0.05}$	$3.17_{\pm0.86}$	$1.76_{\pm0.93}$	$0.49_{\pm0.21}$	$0.07_{\pm0.02}$
	LISA	FT-Real	Good	**0.135**	$0.86_{\pm0.07}$	$3.20_{\pm1.05}$	$1.75_{\pm0.93}$	$0.48_{\pm0.30}$	$0.08_{\pm0.04}$
	LISA	Synth	All	**0.129**	$0.86_{\pm0.04}$	$3.04_{\pm0.69}$	$1.64_{\pm0.69}$	$0.48_{\pm0.21}$	$0.09_{\pm0.03}$
	LISA	Synth	Good	0.236	$0.84_{\pm0.07}$	$3.17_{\pm0.96}$	$1.89_{\pm1.00}$	$0.53_{\pm0.29}$	$0.09_{\pm0.04}$
	LISA$_+$	FT-Real	All	**0.182**	$0.85_{\pm0.06}$	$3.20_{\pm0.99}$	$1.76_{\pm0.98}$	$0.49_{\pm0.26}$	$0.09_{\pm0.03}$
	LISA$_+$	FT-Real	Good	0.196	$0.85_{\pm0.07}$	$3.28_{\pm1.19}$	$1.89_{\pm1.05}$	$0.52_{\pm0.29}$	$0.08_{\pm0.03}$

Model Pre-Training. Using the generated data and corresponding labels, we pre-trained two model variants on the dHCP and BOBs datasets: **(i)** Synth: a model trained only on synthetic data, and **(ii)** FT-Real: a model trained on synthetic data and fine-tuned on the real images from the pre-training dataset, as this was found to yield improved performance in previous work [15]. We did not leverage a mix of real and synthetic data in training, as this approach generally worsens performance by going against the tested philosophy of domain randomization: maximizing training variability to be robust to changes at deployment time. The fine-tuning step also included the randomized resolution resampling and ULF augmentations mentioned above. The training pipeline is summarized in Fig. 2.

Model Fine-Tuning. The pre-trained models were then fine-tuned on the LISA dataset, on all target labels at once. When pre-trained models were segmenting the entire brain, we fine-tuned the model by only optimizing the output channels relevant to our tasks (hippocampi and basal ganglia labels). This step followed the same training parameters as above, with the only exception that we did not use randomized resolution resampling here. After fine-tuning, we also explored various model combinations using ensembling through majority voting [24].

Model Architecture and Optimization Parameters. All individual models we train in our experiments were based on a 3D U-Net implemented in MONAI [25]. The network started with 32 feature channels and doubled the number of channels at each downsampling stage. Convolutional layers used $3 \times 3 \times 3$ kernels with LeakyReLU activations, and the final layer applied a softmax function. Skip connections were included between corresponding encoder and decoder levels. Training was performed with the Adam optimizer (learning rate 10^{-3}) using a combined Dice-Cross-Entropy loss. To ensure stable convergence, we employed a `ReduceLROnPlateau` scheduler (factor 0.1, patience $= 10$) and early stopping (patience $= 100$ iterations). During training, we reserved 10% of the available training data as an internal validation set, which was used to monitor performance and trigger early stopping. All experiments were run on NVIDIA RTX 6000 GPUs using PyTorch Lightning with a batch size of 1. ULF augmentations were generated using TorchIO [23]. Upon acceptance of the paper, we will release the full training and image-generation code as well as the best model weights for reproducibility.

2.4   Experiment Setting

We conducted several rounds of experiments to evaluate our hypotheses. All models were assessed on the LISA challenge validation dataset, and we reported the official metrics published on the challenge website on the day of the paper submission. In all experiments, the models were optimized to segment the whole set of 8 labels provided in the LISA challenge, thus enabling a joint optimization for both sub-tasks 2a and 2b simultaneously.

Metrics. We reported the metrics included in the evaluation of LISA challenge [4], namely Dice Score (DSC; ↑), Hausdorff distance (HD; ↓), 95% HD (HD95; ↓), average symmetric surface distance (ASSD; ↓), and relative volume error (RVE; ↓). The metrics covered only the labels evaluated in the scope of the challenge (e.g., excluding lateral ventricles). In addition, we also computed the *final challenge ranking metric*, which was computed as the mean of the normalized scores of all other metrics[2], each normalized between the worst and best values across all submissions. For simplicity, we refer to this aggregated normalized averaged metric as `NormAvg` throughout the results.

Experiments. Our experiments explored three questions.

1. Can DR help leverage external HF MRI datasets of infants/ neonates for ULF segmentation?

For each combination of pre-training dataset (BOBs or dHCP) and annotation scheme (LISA or LISA$_+$), we trained and evaluated the `Synth` and `FT-Real` variants of our models. We then carried out a qualitative evaluation to assess the out-of-the-box performance of our models pre-trained on the high-field dHCP and BOBs datasets. During this step, we observed some misregistered labels in the LISA data. This prompted us to carry out a manual *quality assessment*: we looked at the right ventricle and caudate nucleus across the data and rated as **bad** the samples where these structures were misregistered during label propagation from HF segmentations.

2. What is the Most Efficient Pre-training For Each Task? In this experiment, we fine-tuned the different pre-trained models of experiment 1 on the LISA dataset and evaluated them quantitatively. We also explored how label quality impacted the fine-tuning performance of these models, using either `all` data, only `good` data, or only misregistered data (denoted as **bad** for conciseness).

3. How to Get the Most Out of Ensembling? Finally, we explored the combination of various fine-tunings of our pre-trained models with different annotation schemes, through a voxel-wise majority voting.

3   Results and Discussion

3.1   Domain Randomization Bridges HF and ULF Annotations

Table 2 summarizes the results of pre-training the models on dHCP and BOBs datasets and their direct evaluation on LISA (no training on LISA data).

Across most experiments, `FT-Real` models (trained on synthetic data and then fine-tuned on real high-field images) consistently outperformed those trained on synthetic data alone. Despite the HF-ULF domain shift, the models benefited from additional T2 information in the HF data. Having never seen the LISA data, these models were not competitive quantitatively, but BOBs-based models still yielded qualitatively good segmentations, especially in the ventricles and caudate nuclei, as illustrated in Fig. 3. This figure also helped us realize that some of the ground truth annotations were *misregistered* (see yellow arrows on the left side of the image). We also observed that the segmentations produced by our pre-trained models were well aligned in these cases.

---

[2] Using 1-DSC to reverse the metric value and align its scale to others.

A manual inspection of the provided training dataset revealed that the HF annotations had some misalignment around the right hemisphere ventricle and caudate in at least 23 out of 79 training cases, similar to those errors depicted in the Fig. 3 A. On Fig. 3 B, the Dice score computed on these structures using our pre-trained models and the ground truth LISA labels did show a correlation between a lower Dice score and data that we rated as misaligned. A more detailed explanation of our manual label quality rating is provided in Appendix S1.

3.2   The Effect of Annotation Quality on Model Training

To assess how annotation quality influences model performance, we fine-tuned the three best-performing pre-trained models (namely BOBs–LISA$_+$–FT-Real, BOBs–LISA–FT-Real, and BOBs–LISA–Synth) on three different subsets of the LISA training data: only *good* annotations (based on our manual review) or the full set of *all* annotations. *Bad* annotations systematically led to poorer scores (NormAvg around 0.43–0.49) and were not reported. The results are summarized in Table 3 and illustrated in Fig. 4. Training on the subset with only good annotations consistently improved performance across both tasks and for nearly all model configurations. For example, the best single model for Task 2a (BOBs-LISA$_+$-FT-Real) improved its NormAvg from 1.24 after pre-training to 0.118 after fine-tuning on only good images. A similar trend is observed in Task 2b, where the BOBs+LISA+Synth model improved from a NormAvg of 1.088 to 0.129 after fine-tuning on good annotations. Generally, models fine-tuned on only good annotations produced segmentations much more closely aligned with anatomical image features, as illustrated in Fig. 4.

3.3   Model Ensembling

In our final set of experiments, we explored various ensembling strategies. For these experiments, we selected the three best models trained on the BOBs dataset with LISA annotations for each task. We also leveraged the two sets of available annotations for the hippocampi and basal ganglia, namely the ones done on ULF (GT_{ULF}) and the ones done on HF and propagated to ULF (GT_{HF}), used for the evaluation. The recipes of models used are presented in Table 4. **M1** and **M2** contain the top-three individual models for tasks 2a and 2b. **M3** and **M4** were built less systematically, through trial of various combinations.

Results are provided in Table 5. Ensembling improves performance compared to individual models. M3 and M4 achieve the strongest performances on Task 2a and 3b, respectively, outperforming the more homogeneous approaches of M1 and M2. M3 and M4, respectively, achieved a NormAvg of 0.0754 for Task 2a and of 0.0319 for Task 2b. We also observed that models trained using GT_{LF} annotations consistently outperformed those trained on GT_{HF} annotations, even when evaluated on them. We believe that this could warrant further investigation for a future edition of the challenge.

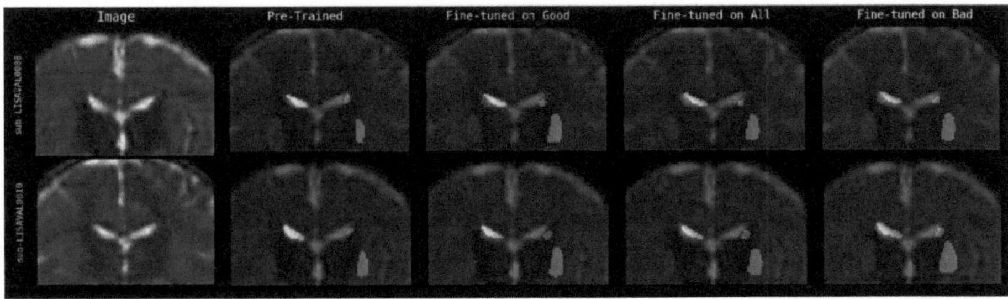

Fig. 4. Effect of annotation quality on segmentation results on the BOBs–LISA model. Fine-tuning with only good annotations leads to clearly improved alignment with image features, whereas fine-tuning with bad or mixed annotations introduces misalignment due to registration errors, which are absent on the segmentations of the pre-trained model.

Table 4. Recipes used for model ensembling. **Type** refers to the quality type of annotation used for fine-tuning (**all** samples or only the ones rated **good**), and **GT** refers to the labeling scheme used for the LISA data.

Merged model	Model 1			Model 2			Model 3		
	Pre-training	Type	GT	Pre-training	Type	GT	Pre-training	Type	GT
M1	BOBs+LISA+Synth	all	GT_{HF}	BOBs+LISA+FT-Real	good	GT_{HF}	BOBs+LISA+Synth	good	GT_{HF}
M2	BOBs+LISA+Synth	all	GT_{HF}	BOBs+LISA+FT-Real	good	GT_{HF}	BOBs+LISA+Synth	good	GT_{HF}
M3	dHCP+LISA$_+$+Synth	all	GT_{LF}	dHCP+Synth+LISA$_+$	good	GT_{LF}	BOBs+LISA$_+$+FT-Real	good	GT_{HF}
M4	— Model **M1** —			— Model **M2** —			dHCP+LISA$_+$+Synth	all	GT_{LF}

4   Discussion

This study shows that domain randomization, combined with careful data curation and model fusion, can effectively bridge the gap between HF and ULF MRI for neonatal brain segmentation and, in some cases, work better for annotation transfer from HF to ULF than simple subject co-registration approaches.

Domain randomization significantly improves generalization: BOBs-pretrained models allowed to achieve strong transfer performance when fine-tuned on LISA data. Both LISA and LISA$_+$ annotations were useful, with LISA$_+$ performing best on Task 2a and LISA-annotations on Task 2b.

An important advantage of domain randomization is that, since images are generated directly from segmentation maps, the resulting image-annotation pairs are perfectly aligned, with image contrasts matching annotation borders – something difficult to achieve with manual annotations. Annotation quality proved critical: models trained on poorly aligned annotations overfit to registration errors, while using only high-quality annotations improved performance and alignment with image features. HF annotations transferred to LF images via registration were unreliable in around 30% of the data, and excluding them from training helped improve performance.

Table 5. Comparison of ensembling strategies defined in Table 4 and best individual models (resp. BOBs+LISA$_+$+**FT-Real** and BOBs+LISA+**Synth**).

Task	Strategy	NormAvg ↓	DSC ↑	HD ↓	HD95 ↓	ASSD ↓	RVE ↓
2A–Hippo.	Best indiv. 2a	0.118	$0.69_{\pm0.18}$	$5.73_{\pm7.79}$	$2.02_{\pm1.42}$	$0.80_{\pm0.90}$	$0.12_{\pm0.08}$
	Best indiv. 2b	0.183	$0.70_{\pm0.19}$	$6.06_{\pm7.74}$	$2.23_{\pm1.56}$	$0.81_{\pm0.95}$	$0.18_{\pm0.07}$
	M1	0.141	$0.70_{\pm0.19}$	$5.79_{\pm7.91}$	$2.11_{\pm1.41}$	$0.77_{\pm0.88}$	$0.16_{\pm0.08}$
	M2	0.161	$0.69_{\pm0.19}$	$6.08_{\pm8.06}$	$2.15_{\pm1.48}$	$0.80_{\pm0.94}$	$0.14_{\pm0.08}$
	M3	**0.075**	$0.71_{\pm0.19}$	$5.65_{\pm7.98}$	$1.95_{\pm1.49}$	$0.76_{\pm0.96}$	$0.13_{\pm0.07}$
	M4	0.125	$0.70_{\pm0.19}$	$5.90_{\pm7.94}$	$2.07_{\pm1.48}$	$0.77_{\pm0.89}$	$0.14_{\pm0.07}$
2B–BaGA	Best indiv. 2a	0.196	$0.85_{\pm0.07}$	$3.28_{\pm1.19}$	$1.89_{\pm1.05}$	$0.52_{\pm0.29}$	$0.08_{\pm0.03}$
	Best indiv. 2b	0.129	$0.86_{\pm0.04}$	$3.04_{\pm0.69}$	$1.64_{\pm0.69}$	$0.48_{\pm0.21}$	$0.09_{\pm0.03}$
	M1	0.076	$0.86_{\pm0.05}$	$2.92_{\pm0.84}$	$1.59_{\pm0.79}$	$0.45_{\pm0.24}$	$0.07_{\pm0.02}$
	M2	0.065	$0.86_{\pm0.05}$	$3.02_{\pm0.86}$	$1.61_{\pm0.83}$	$0.44_{\pm0.23}$	$0.06_{\pm0.03}$
	M3	0.093	$0.86_{\pm0.06}$	$2.96_{\pm1.10}$	$1.69_{\pm0.91}$	$0.46_{\pm0.26}$	$0.07_{\pm0.03}$
	M4	**0.032**	$0.87_{\pm0.05}$	$3.01_{\pm0.82}$	$1.56_{\pm0.77}$	$0.44_{\pm0.22}$	$0.06_{\pm0.03}$

Finally, model ensembling further boosted performance. As commonly observed in medical imaging challenges [24], a highly diverse, heuristically built ensemble of models trained throughout our experiments was found to achieve the best performance, although this approach might not generalize as well to new tasks as more principled ensembles.

In this work, we illustrated the strength of domain-randomization-based approaches for HF-to-ULF knowledge transfer. We believe that these methods could be instrumental in designing robust and generalizable models for ULF pediatric segmentation tasks.

Acknowledgments. This research was funded by the Swiss National Science Foundation (215641), ERA-NET Neuron MULTI-FACT project (SNSF 31NE30_203977); we acknowledge the Leenaards and Jeantet Foundations as well as CIBM Center for Biomedical Imaging, a Swiss research center of excellence founded and supported by CHUV, UNIL, EPFL, UNIGE and HUG. This research was also supported by grants from NVIDIA and utilized NVIDIA RTX6000 ADA GPUs. The Developing Human Connectome Project (dHCP) was funded by the European Research Council (ERC) under the European Union's Seventh Framework Programme (FP7/2007–2013), Grant Agreement No. 319456.

Disclosure of Interests. The authors have no competing interests to disclose.

References

1. Barnea-Goraly, N., et al.: A preliminary longitudinal volumetric mri study of amygdala and hippocampal volumes in autism. Prog. Neuropsychopharmacol. Biol. Psychiatry **48**, 124–128 (2014)
2. Hoogman, M., et al.: Subcortical brain volume differences in participants with attention deficit hyperactivity disorder in children and adults: a cross-sectional mega-analysis. Lancet Psychiatry **4**(4), 310–319 (2017)
3. Xu, Q., Zuo, C., Liao, S., Long, Y., Wang, Y.: Abnormal development pattern of the amygdala and hippocampus from childhood to adulthood with autism. J. Clin. Neurosci. **78**, 327–332 (2020)
4. Lepore, N., et al.: Low field pediatric brain magnetic resonance image segmentation and quality assurance (lisa) (2025). https://doi.org/10.5281/zenodo.15081583
5. Jalloul, M., et al.: Mri scarcity in low-and middle-income countries. NMR Biomed. **36**(12), e5022 (2023)
6. Murali, S., et al.: Bringing mri to low-and middle-income countries: directions, challenges and potential solutions. NMR Biomed. **37**(7), e4992 (2024)
7. Iglesias, J.E., et al.: Quantitative brain morphometry of portable low-field-strength mri using super-resolution machine learning. Radiology, **306**(3) (2023). ISSN 1527-1315. https://doi.org/10.1148/radiol.220522
8. Zhao, Y., et al.: Whole-body magnetic resonance imaging at 0.05 tesla. Science, **384**(6696), eadm7168 (2024)
9. Johnson, I., et al.: Automated segmentation of white matter hyperintensities on portable low-field magnetic resonance imaging (p9-13.002). Neurology, **104** (2025). ISSN 1526-632X
10. Gopinath, K., et al. From low field to high value: robust cortical mapping from low-field mri. arXiv preprint arXiv:2505.12228 (2025)
11. Billot, B., et al.: Synthseg: Segmentation of brain mri scans of any contrast and resolution without retraining. Med. Image Anal. **86**, 102789 (2023)
12. Billot, B., Magdamo, C., Cheng, Y., Arnold, S.E., Das, S., Iglesias, J.E.: Robust machine learning segmentation for large-scale analysis of heterogeneous clinical brain mri datasets. Proc. National Academy Sci. **120**(9), e2216399120 (2023)
13. Váša, F., et al. Ultra-low-field brain MRI morphometry: test-retest reliability and correspondence to high-field MRI. BioRxiv, 2024–08 (2024)
14. Valabregue, R., Girka, F., Pron, A., Rousseau, F., Auzias, G.: Comprehensive analysis of synthetic learning applied to neonatal brain mri segmentation. Human Brain Mapp. **45**(6), e26674 (2024)
15. Zalevskyi, V., et al.: Maximizing domain generalization in fetal brain tissue segmentation: the role of synthetic data generation, intensity clustering and real image fine-tuning. arXiv preprint arXiv:2411.06842 (2024)
16. A. David Edwards, et al.: The developing human connectome project neonatal data release. Front. Neurosci., **16** (2022). ISSN 1662-453X. https://doi.org/10.3389/fnins.2022.886772
17. Feczko, E., et al.: Baby open brains: an open-source repository of infant brain segmentations (2024). https://doi.org/10.1101/2024.10.02.616147
18. Feczko, E., et al.: Baby open brains: an open-source repository of infant brain segmentations. bioRxiv (2024)
19. Howell, B.R., et al.: The UNC/UMN baby connectome project (BCP): an overview of the study design and protocol development. NeuroImage, **185**, 891–905 (2019). ISSN 1053-8119. https://doi.org/10.1016/j.neuroimage.2018.03.049

20. Edwards, A.D., et al.: The developing human connectome project neonatal data release. Front. Neurosci. **16**, 886772 (2022)
21. Makropoulos, A., et al.: The developing human connectome project: a minimal processing pipeline for neonatal cortical surface reconstruction. NeuroImage, **173**, 88–112 (2018). ISSN 1053-8119. https://doi.org/10.1016/j.neuroimage.2018.01.054
22. Deoni, S.C.L., O'Muircheartaigh, J., Ljungberg, E., Huentelman, M., Williams, S.C.: Simultaneous high-resolution t2-weighted imaging and quantitative t 2 mapping at low magnetic field strengths using a multiple te and multi-orientation acquisition approach. Magnetic Res. Med. **88**(3), 1273–1281 (2022)
23. Pérez-García, F., Sparks, R., Ourselin, S.: Torchio: a python library for efficient loading, preprocessing, augmentation and patch-based sampling of medical images in deep learning. Comput. Methods Programs Biomed. **208**, 106236 (2021)
24. Eisenmann, M., et al.: Why is the winner the best? In: Proceedings of the IEEE/CVF Conference on Computer Vision and Pattern Recognition, pp. 19955–19966 (2023)
25. MONAI Consortium. Monai: Medical open network for AI (2025)

Enforcing Anatomical Symmetry with Euclidean Distance Transforms for Low-Field MRI Bilateral Structure Segmentation

Zdravko Marinov[1]([envelope]) [iD], Jens Kleesiek[2,3] [iD], and Rainer Stiefelhagen[1] [iD]

[1] Karlsruhe Institute of Technology, Karlsruhe, Germany
`zdravko.marinov@kit.edu`
[2] Institute for AI in Medicine,University Hospital Essen, Essen, Germany
[3] Cancer Research Center Cologne Essen (CCCE), University Medicine Essen, Essen, Germany

Abstract. Accurate segmentation of subcortical brain structures in MRI is essential for the study of neurodevelopment, particularly in pediatric populations. While low-field MRI scanners offer a cost-effective and safer alternative to high-field systems—especially eliminating the need for sedation in young children—they present challenges due to lower image resolution and signal-to-noise ratio. In this work, we propose a symmetry-aware post-processing strategy to improve the segmentation of bilateral structures in low-field MRI. We first train baseline U-Net models for the segmentation of eight anatomical structures, including hippocampi, in the LISA 2025 pediatric low-field MRI dataset. While these models achieve reasonable accuracy, we observe frequent violations of anatomical symmetry in their predictions. To address this, we introduce a novel correction step that explicitly enforces plausible anatomical symmetry by identifying discrepancies between hemispheres and applying deformation fields anchored by the dominant structure from each symmetric pair. This post-hoc alignment improves segmentation quality for all symmetric targets, particularly the hippocampi. Our approach highlights the importance of leveraging anatomical priors in low-resource imaging scenarios and paves the way for more reliable analyses in global health contexts(Code: https://github.com/Zrrr1997/LISA_2025_cvhci).

Keywords: Hippocampi Segmentation · Low-field MRI · Anatomical Symmetry · Post-processing

1 Introduction

Accurate segmentation of subcortical brain structures in neonatal magnetic resonance imaging (MRI) plays a critical role in studying early brain development [1], identifying markers of neurodevelopmental disorders [4], and enabling

N. Lepore and M. G. Linguraru (Eds.): LISA 2025, LNCS 16411, pp. 63–73, 2026.
https://doi.org/10.1007/978-3-032-14417-1_6

population-level analyses of anatomical variability [8]. Among these structures, the *hippocampus* is especially important for memory formation, learning, and spatial navigation [10,17]. The *basal ganglia*—including the caudate and lentiform nuclei—are involved in motor control, reward processing, and cognitive function [5,9], while the *lateral ventricles* are relevant for monitoring cerebrospinal fluid (CSF) dynamics and detecting abnormal enlargement, such as in hydrocephalus [14,16].

Although high-field (HF) MRI systems (e.g., 3T and 7T scanners) are the gold standard for neonatal brain imaging due to their high spatial resolution and contrast, their use is limited by high costs, the need for sedation, and restricted availability in low-resource or rural settings. Recently, portable low-field (LF) MRI systems, such as HyperfineâĂŹs Swoop scanner, have emerged as promising alternatives. These systems are safer, more affordable, and easier to deploy in under-resourced hospitals, making them well-suited for neonatal care. However, their lower signal-to-noise ratio and reduced spatial fidelity pose significant challenges for automated image analysis, particularly for fine-grained segmentation tasks.

The LISA 2025 challenge provides a comprehensive neonatal brain MRI dataset featuring Low-Field (LF) T2-weighted scans acquired from multiple international sites, including Makerere University Hospital (Uganda), University of Cape Town (South Africa), and Rhode Island Hospital (USA). Each subject underwent imaging with both high-field (HF) and low-field MRI systems, with the LF scans rigidly registered to their corresponding HF reference frames via a 9-point linear registration. The dataset includes expertly annotated HF segmentations of bilateral hippocampi and basal ganglia structures—specifically the caudate and lentiform nuclei—which serve as the primary targets for segmentation. While manual segmentations of the lateral ventricles in LF space are also provided, these are intended for auxiliary analyses and are not included in the official evaluation metrics. Participants were tasked with generating accurate segmentations of bilateral hippocampi and basal ganglia directly from the registered LF images, following a standardized labeling scheme aligned with the provided ground truths.

2   Methods

2.1   U-Net Training

In our study, we train an ensemble of 5 U-Net models [15] via 5-fold cross-validation for the automatic segmentation of eight key subcortical anatomical structures, categorized as follows:

- **Bilateral hippocampi:** left (label 1), right (label 2),
- **Bilateral lateral ventricles:** left (label 3), right (label 4)[1],
- **Bilateral caudate nuclei:** left (label 5), right (label 6),
- **Bilateral lentiform nuclei:** left (label 7), right (label 8).

---

[1] Ventricles are provided as auxiliary ground truth but are not part of the official evaluation.

We also train a second dedicated ensemble of 5 models only on the task of hippocampus segmentation for labels 1 and 2. During our model development, we observed that the predicted segmentations frequently display asymmetric representations of bilaterally symmetric structures, with the strongest asymmetry seen in the hippocampi (see Fig. 1). These asymmetries can undermine anatomical plausibility and affect downstream clinical or research interpretations, especially in a neonatal context where such structures are expected to exhibit a certain degree of anatomical symmetry. In the end, we ensemble all 10 trained models for the hippocampus segmentation Task2a and the 5 models trained on all structures for Task2b via majority voting.

Architecture and Training Details. We use the SW-FastEdit [6] implementation of U-Net for our submission, which was originally used for interactive click-based PET/CT lesion annotation [12,13] on the autoPET dataset [2,3]. However, we apply its non-interactive (without clicks) implementation [7] and adapt it to low-field MRI images by simply adjusting its pre-processing transforms and keeping the same training scheme. The training is performed using a sliding window inferer with a patch size of $[128, 128, 128]$ and overlap of 25% with Gaussian weighting. We compute the predictions over the whole training image and then average the gradients over the overlapping regions equally to compute the final loss. We also utilize a cosine annealing learning rate scheduler with an initial learning rate of 1e-4 and train all of our models for 50 epochs without any data augmentation. We normalize each MRI image with a standard z-score normalization transform. We use the Dice Loss combined with the cross-entropy loss, weighted equally. We utilize the MONAI DynUNet[2] backbone with only three encoder-decoder levels, leading to very small and efficient models of only slightly over 1 million learnable parameters. The training parameters are summarized in Table 1.

2.2   Post-Processing for Anatomical Symmetry

To promote biologically plausible and anatomically consistent segmentations, particularly for brain structures that are expected to exhibit bilateral symmetry, we introduce a dedicated post-processing step designed to reduce volume imbalances between corresponding left-right anatomical regions. This symmetry harmonization step specifically targets pairs of structures such as the left and right hippocampi or the bilateral components of the basal ganglia, where natural symmetry is a reasonable assumption in healthy neonatal brains.

The procedure comprises two key stages. First, for each bilateral structure, we retain only the largest connected component (LCC) in the predicted mask. This step serves to eliminate small, isolated false positives or noise that may have been introduced during segmentation, particularly in low signal-to-noise regions of the low-field MRI scans. Second, we iteratively grow the smaller of

---

[2] https://docs.monai.io/en/0.7.0/_modules/monai/networks/nets/dynunet.html

Table 1. Summary of training parameters used for U-Net segmentation.

Parameter	Value
Architecture	MONAI DynUNet (3 encoder-decoder levels)
Parameter Count	≈ 1 million
Backbone Source	SW-FastEdit [6]
Input Patch Size	$[128, 128, 128]$
Sliding Window Overlap	25%
Window Weighting	Gaussian
Gradient Computation	Whole-image with overlapping region gradient averaging
Loss Function	Dice + Cross-Entropy (equal weights)
Learning Rate	1×10^{-4} (cosine annealing schedule)
Epochs	50
Data Augmentation	None
Normalization	Z-score (per image)
Training Mode	Non-interactive (no clicks) [7]

the two structures using morphological operations guided by distance transform maps, until the volumetric discrepancy between the pair falls below a predefined tolerance threshold, denoted as τ. In essence, we aim to equalize the volumes within a specified margin, under the assumption that the underlying anatomy should not exhibit drastic asymmetry unless pathological.

This process is formally described in Algorithm 1. For Task 2a (hippocampus segmentation), we empirically set $\tau = 10\%$, meaning that if one hippocampus is more than 10% larger than its contralateral counterpart, the algorithm will attempt to iteratively grow the smaller one until the relative volume difference is reduced below that margin. For Task 2b (basal ganglia segmentation), we apply the same logic with a slightly stricter threshold of $\tau = 8\%$, based on the observed variability in training data. These thresholds were chosen after analyzing the training set and identifying the maximum naturally occurring asymmetries between paired structures in ground truth annotations. Our goal was to enforce symmetry only in cases where the modelâĂŹs predictions deviated beyond what would be considered biologically typical.

Clarification on Algorithm 1: We compute the largest connected component for each label (right and left) and then check if the ratio between the volumes of these components is larger than our threshold τ. If that is the case, we compute the Euclidean distance transform (EDT) for each component and flip the right EDT on top of the left one. We then check, for each voxel, which of the two distance fields has the greater value and define the target thickness map T as the voxel-wise maximum between the left and mirrored-right EDTs. Intuitively, this map captures the desired extent of the structure on each side assuming perfect symmetry. Next, we identify the side with the smaller volume (denoted as A)

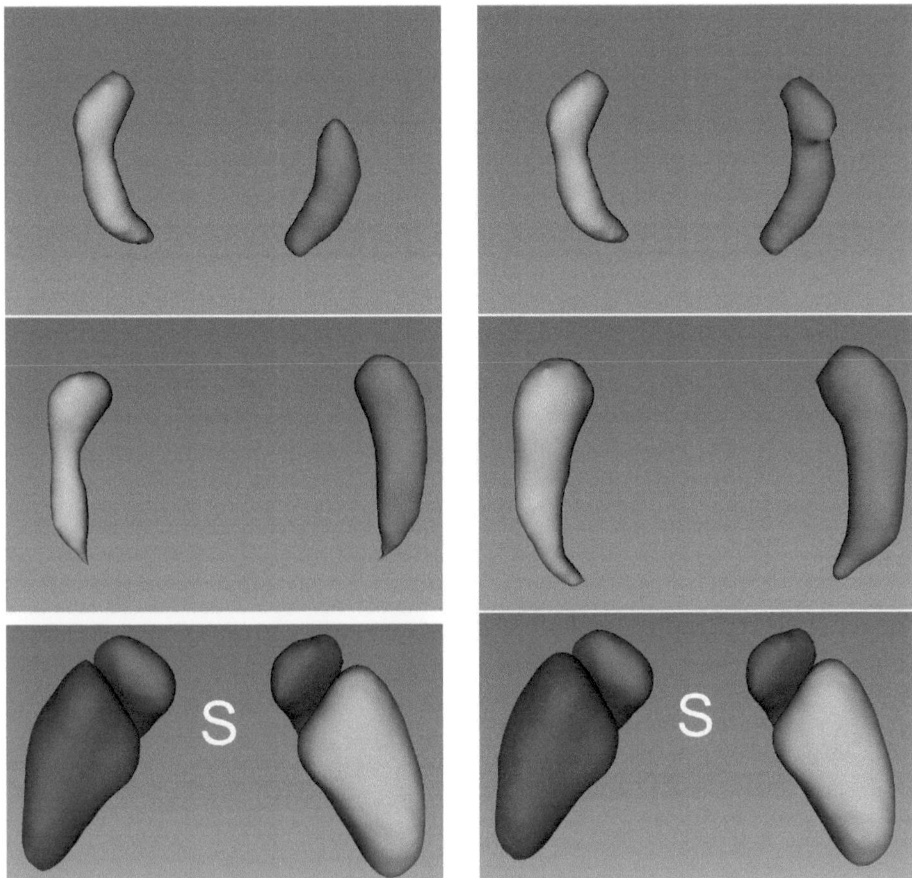

Fig. 1. (Left) Typical segmentations for two samples from the validationion set of our trained models that exhibit a large asymmetry between the left and right hippocampi. (Right) Result after applying our symmetry-enforcing post-processing.

and define its *growth zone* as the set of voxels outside the current component ($\neg mask_A$) but within the target thickness T. This ensures that growth only occurs where the current component is thinner than its mirrored counterpart. We then iteratively expand the smaller component by performing a binary dilation restricted to the growth zone, effectively adding voxels that make the shape more symmetric. This process continues until either (i) the relative volume difference between both sides falls below the threshold τ, or (ii) the maximum number of iterations N is reached. The resulting segmentation mask $\tilde{M}$ thus enforces volumetric and shape symmetry between corresponding anatomical structures while preserving their spatial correspondence via the mirrored distance fields (Fig. 2).

Algorithm 1: Symmetry Harmonization via Distance Transforms

---

1 **Input**: Prediction mask M, left label l_1, right label l_2, threshold τ, max iterations N

2 **Output**: Updated segmentation mask $\tilde{M}$

3 $\tilde{M} \leftarrow M$

4 $\tilde{M} \leftarrow \text{keep\_largest\_cc}(\tilde{M}, l_1), \text{keep\_largest\_cc}(\tilde{M}, l_2)$

5 **for** $n = 1$ *to* N **do**

6     $V_1 \leftarrow \text{volume}(\tilde{M} == l_1)$

7     $V_2 \leftarrow \text{volume}(\tilde{M} == l_2)$

8   **if** $\frac{|V_1 - V_2|}{\max(V_1, V_2)} \leq \tau$ **then**

9   **break**

10   **if** $V_1 < V_2$ **then**

11         $A \leftarrow l_1,\ B \leftarrow l_2$

12   **else**

13         $A \leftarrow l_2,\ B \leftarrow l_1$

14     $mask_A \leftarrow (\tilde{M} == A)$

15     $mask_B \leftarrow (\tilde{M} == B)$

16     $DT_A \leftarrow \text{distance\_transform}(mask_A)$

17     $DT_B \leftarrow \text{distance\_transform}(mask_B)$

18     $DT_B^{\text{flip}} \leftarrow \text{flip}(DT_B, \text{left-right axis})$

19     $T \leftarrow \max(DT_A, DT_B^{\text{flip}})$ # target thickness

20     $growth\_zone \leftarrow (\neg mask_A) \wedge (DT_A < T)$

21     $dilated \leftarrow \text{binary\_dilation}(mask_A)$

22     $new\_voxels \leftarrow dilated \wedge growth\_zone$

23     $\tilde{M}[new\_voxels] \leftarrow A$

24 **return** $\tilde{M}$

---

3   Results

3.1   Cross-Validation Results

We performed 5-fold cross-validation to estimate the model performance on both tasks. The results are summarized in Tables 2 and 3. For Task 2a (hippocampus segmentation), the Dice Similarity Coefficient (DSC) values vary notably across folds, reflecting some variability in how well the model generalizes across the different subsets of the data. Before post-processing, the average DSC ranges from 0.55 to 0.65, with a relatively high standard deviation, suggesting that segmentations are sometimes inconsistent or affected by noise—likely due to the small size and low contrast of the hippocampi in low-field MRI.

After applying our post-processing step—which enforces symmetry and removes small disconnected predictions—the scores show slight improvements or stabilization across most folds. While the mean values donâĂŹt jump dramatically (e.g., $0.62 \rightarrow 0.63$), we do observe a consistent reduction in standard deviation, indicating more stable and reliable segmentations across subjects.

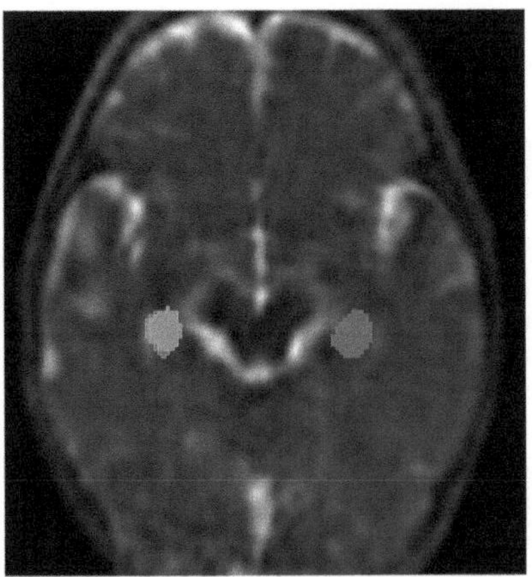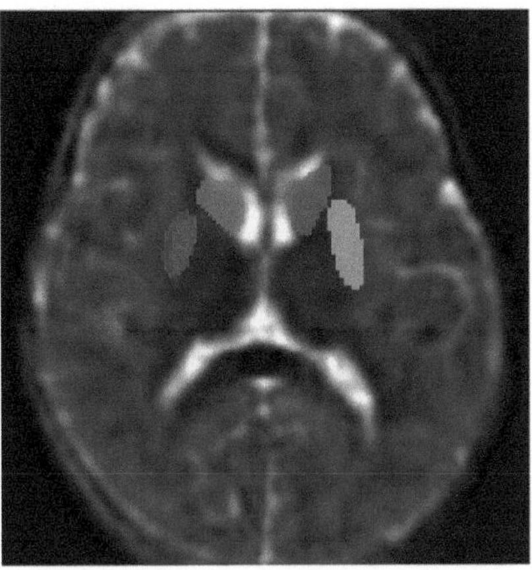

Fig. 2. (Left) Typical example of how our post-processing expands the smaller hippocampus and (Right) the smaller structures from Task2b.

Importantly, both the left and right hippocampus scores improve slightly, suggesting the method does not bias one side over the other.

In Task 2b (basal ganglia segmentation), the results are generally stronger to begin with. The pre-processing DSCs range between 0.78 and 0.82, showing much less variability across folds. This is expected, as the caudate and lentiform nuclei are typically larger and better defined than the hippocampi in neonatal MRI.

The post-processing procedure again helps to slightly tighten performance, with small improvements or stabilization of the mean DSC and lower standard deviations overall. Improvements are subtle, but theyâĂŹre consistent: for example, the left caudate DSC goes from 0.75 to 0.77 in fold 1, and several other structures show similar gains. The results confirm that enforcing anatomical symmetry does not hurt the performance, and in some cases, can boost it by correcting asymmetries or inconsistencies in the raw predictions.

3.2 Results on the Official Validation Set

The results of our ensemble of 10 models for both tasks (10 models for Task 2a and 5 models for Task 2b) is summarized in Table 4. The metrics reported include Dice Similarity Coefficient (DSC), Hausdorff Distance (HD and HD95), Average Symmetric Surface Distance (ASSD), and Relative Volume Error (RVE), offering a comprehensive view of both overlap and boundary accuracy.

Starting with the hippocampi, the average DSC is 0.70 ± 0.17, which reflects moderate segmentation performance with some variability. The left hippocampus shows slightly weaker performance (0.67 DSC) and higher variance, which aligns with the fact that this structure tends to be smaller and harder to segment consistently in low-field neonatal MRI as seen in the results from LISA 2024 [11]. The high standard deviation in HD ($9.52 \pm 15.72\,\mathrm{mm}$) for the left side also

Table 2. Dice Similarity Coefficients (DSC) across cross-validation folds for hippocampus segmentation in Task 2a. For clarity, we only report the standard deviation on the mean DSC.

Metric	Fold				
	0	1	2	3	4
Before Post-processing					
DSC (Mean)	0.64 ± 0.17	0.61 ± 0.19	0.65 ± 0.15	0.62 ± 0.19	0.55 ± 0.17
DSC Left Hipp.	0.64	0.60	0.68	0.60	0.53
DSC Right Hipp.	0.64	0.63	0.62	0.63	0.57
After Post-processing					
DSC (Mean)	0.64 ± 0.16	0.63 ± 0.19	0.66 ± 0.14	0.63 ± 0.17	0.56 ± 0.17
DSC Left Hipp.	0.64	0.61	0.68	0.61	0.55
DSC Right Hipp.	0.64	0.64	0.64	0.63	0.58

Table 3. Dice Similarity Coefficients (DSC) across cross-validation folds for basal ganglia segmentation in Task 2b. For clarity, we only report the standard deviation on the mean DSC.

Metric	Fold				
	0	1	2	3	4
Before Post-processing					
DSC (Mean)	0.81 ± 0.07	0.78 ± 0.07	0.82 ± 0.04	0.80 ± 0.07	0.79 ± 0.07
DSC Left Caud.	0.82	0.75	0.83	0.79	0.76
DSC Right Caud.	0.79	0.78	0.83	0.79	0.78
DSC Left Lenti.	0.82	0.79	0.81	0.80	0.80
DSC Right Lenti.	0.82	0.79	0.81	0.80	0.80
After Post-processing					
DSC (Mean)	0.81 ± 0.05	0.79 ± 0.07	0.82 ± 0.04	0.80 ± 0.05	0.79 ± 0.06
DSC Left Caud.	0.82	0.77	0.83	0.80	0.78
DSC Right Caud.	0.79	0.77	0.83	0.80	0.79
DSC Left Lenti.	0.82	0.81	0.81	0.81	0.80
DSC Right Lenti.	0.82	0.78	0.81	0.80	0.80

indicates that a few difficult cases skew the boundary accuracy, though the more robust HD95 (2.63 ± 1.56 mm) confirms that these outliers are limited.

The right hippocampus is segmented more accurately on average (0.72 DSC), with better boundary alignment (HD = 4.56 mm, ASSD = 0.64 mm), and slightly lower volume error (RVE = 0.19). These differences between sides might stem from asymmetries in image contrast or anatomical variation.

In contrast, performance on the basal ganglia structures is much stronger and more consistent. The caudate nuclei and lentiform nuclei each achieve DSC scores around 0.84âĂŞ0.85, with low standard deviations, indicating the model's reliability across subjects. All related boundary metrics—HD, HD95, and ASSD—are also tighter and more favorable, generally staying below 2 mm on average, with ASSD values around 0.4–0.6 mm. These structures are larger and better defined in T2-weighted scans, which likely contributes to the stronger performance.

Finally, the combined score for all basal ganglia structures shows that the ensemble delivers robust, high-quality segmentations overall (DSC = 0.85 ± 0.05, HD95 = 1.97 mm, ASSD = 0.53 mm, RVE = 0.09). These results highlight both the effectiveness of the ensemble and the relative ease of segmenting basal ganglia compared to smaller, more variable structures like the hippocampi.

Table 4. Evaluation metrics averaged across the official validation set.

Structure	DSC	HD	HD95	ASSD	RVE
Left Hippocampus	0.67±0.22	9.52±15.72	2.63±1.56	0.89±0.89	0.22±0.20
Right Hippocampus	0.72±0.13	4.56±1.34	2.30±0.76	0.64±0.33	0.19±0.16
Both Hippocampi	0.70±0.17	7.04±7.79	2.47±1.10	0.76±0.60	0.20±0.17
Caudate Left	0.84±0.06	3.76±1.33	1.70±0.93	0.45±0.21	0.07±0.05
Caudate Right	0.85±0.07	4.47±2.07	1.89±1.00	0.44±0.26	0.08±0.05
Lentiform Left	0.85±0.05	3.63±0.95	2.27±0.95	0.63±0.24	0.10±0.08
Lentiform Right	0.85±0.05	3.34±1.06	2.02±0.95	0.59±0.26	0.10±0.11
All Basal Ganglia	0.85±0.05	3.80±1.04	1.97±0.89	0.53±0.21	0.09±0.05

4 Discussion

This study tackled the challenge of segmenting subcortical brain structures in low-field pediatric MRI, where low resolution and contrast make accurate segmentation difficult. Our U-Net ensemble performed well overall, especially for the basal ganglia, but hippocampus segmentation was more variable and prone to errors—likely due to its small size and unclear boundaries in low-field images.

A recurring issue was anatomically implausible asymmetry between predicted left and right structures, particularly the hippocampi. To address this, we introduced a symmetry-aware post-processing step that harmonizes structure volumes across hemispheres by identifying and correcting substantial discrepancies. This approach is grounded in a simple anatomical prior: that bilateral structures in the brain tend to be symmetric, especially in healthy pediatric populations.

Post-processing led to small but consistent improvements in Dice scores, and qualitatively improved the anatomical plausibility of segmentations. The gains were more pronounced in Task 2a (hippocampi) than in Task 2b (basal ganglia), where symmetry violations were less frequent.

One advantage of our method is its simplicity and flexibility. Rather than modifying the model architecture or training objective, we enforce symmetry in a lightweight, interpretable post-hoc step. This is especially useful in low-resource settings where robust, efficient tools are needed. However, the method assumes that the larger of each bilateral pair is more accurate, which may not always hold—particularly in pathological cases. Future work could integrate uncertainty estimation or learn symmetry constraints directly during training.

Our approach shows that incorporating anatomical priors—even in a basic form—can meaningfully enhance segmentation quality in low-quality imaging scenarios. This has promising implications for making neuroimaging tools more reliable and accessible in global health and pediatric research.

Acknowledgments. The present contribution is supported by the Helmholtz Association under the joint research school "HIDSS4Health âĂŞ Helmholtz Information and Data Science School for Health. Parts of this work were performed on the HoreKa supercomputer, funded by the Ministry of Science, Research, and the Arts of Baden-WÃijrttemberg, and by the Federal Ministry of Education and Research.

References

1. Dubois, J., Dehaene-Lambertz, G., Kulikova, S., Poupon, C., Hüppi, P., Hertz-Pannier, L.: The early development of brain white matter: a review of imaging studies in fetuses, newborns and infants. Neuroscience **276**, 48–71 (2014)
2. Gatidis, S., et al.: The autopet challenge: towards fully automated lesion segmentation in oncologic PET/CT imaging (2023)
3. Gatidis, S., et al.: Results from the autopet challenge on fully automated lesion segmentation in oncologic PET/CT imaging. Nat. Mach. Intell. **6**(11), 1396–1405 (2024)
4. Gilmore, J., Lin, W., Prastawa, M., et al.: Regional gray matter growth, sexual dimorphism, and cerebral asymmetry in the neonatal brain. J. Neurosci. **27**(6), 1255–1260 (2007)
5. Graybiel, A.: The basal ganglia. Curr. Biol. **10**(14), R509–R511 (2000)
6. Hadlich, M., Marinov, Z., Kim, M., Nasca, E., Kleesiek, J., Stiefelhagen, R.: Sliding window FASTEDIT: a framework for lesion annotation in whole-body pet images. In: 2024 IEEE International Symposium on Biomedical Imaging (ISBI), pp. 1–5. IEEE (2024)
7. Hadlich, M., Marinov, Z., Stiefelhagen, R.: Autopet challenge 2023: sliding window-based optimization of u-net. arXiv preprint arXiv:2309.12114 (2023)
8. Kuklisova-Murgasova, M., Aljabar, P., Srinivasan, L., et al.: A dynamic 4d probabilistic atlas of the developing brain. Neuroimage **54**(4), 2750–2763 (2011)
9. Lanciego, J., Luquin, N., Obeso, J.: Functional neuroanatomy of the basal ganglia. Cold Spring Harbor Perspect. Med. **2**(12), a009621 (2012)
10. Lavenex, P., Banta Lavenex, P.: Building hippocampal circuits to learn and remember: Insights into the development of human memory. Behav. Brain Res. **254**, 8–21 (2013)
11. Lepore, N., Linguraru, M.G.: Low Field Pediatric Brain Magnetic Resonance Image Segmentation and Quality Assurance: First MICCAI Challenge, LISA 2024, Held in Conjunction with MICCAI 2024, Marrakesh, Morocco, October 10, 2024, Proceedings. Springer Nature (2025)

12. Marinov, Z., Jäger, P.F., Egger, J., Kleesiek, J., Stiefelhagen, R.: Deep interactive segmentation of medical images: a systematic review and taxonomy. IEEE Trans. Patt. Anal. Mach. Intell. (2024)
13. Marinov, Z., Stiefelhagen, R., Kleesiek, J.: Guiding the guidance: a comparative analysis of user guidance signals for interactive segmentation of volumetric images. In: International Conference on Medical Image Computing and Computer-Assisted Intervention, pp. 637–647. Springer (2023)
14. McAllister, J.: Pathophysiology of congenital and neonatal hydrocephalus. Semi. Fetal Neonatal Med. **17**(5), 285–294 (2012)
15. Ronneberger, O., Fischer, P., Brox, T.: U-net: convolutional networks for biomedical image segmentation. In: International Conference on Medical Image Computing and Computer-Assisted Intervention, pp. 234–241. Springer (2015)
16. Shen, M., Kim, S., McKinstry, R., et al.: Early brain enlargement and elevated extra-axial fluid in infants who develop autism spectrum disorder. Brain **136**(9), 2825–2835 (2013)
17. Squire, L.: Memory and the hippocampus: a synthesis from findings with rats, monkeys, and humans. Psychol. Rev. **99**(2), 195–231 (1992)

Coordinate Transformations Make Segmentation Models More Data-Efficient

Mahbod Issaiy$^{(\boxtimes)}$ [ID]

Advanced Diagnostic and Interventional Radiology Research Center (ADIR), Tehran University of Medical Sciences, Tehran, Iran
mahbodissaiy@gmail.com

Abstract. Ultra-low-field (0.064T) magnetic resonance imaging (MRI) systems enable portable brain imaging but pose significant challenges for automated segmentation due to low signal-to-noise ratio and limited resolution. We present a coordinate transform-based deep learning approach for pediatric brain structure segmentation that analytically handles geometric variability through spherical and ellipsoidal coordinate mappings. Our method employs an ensemble of SwinUNETR models trained in Cartesian, spherical, and ellipsoidal spaces, combined with two novel loss functions: Projection Dice Loss for shape-aware supervision through 2D orthogonal projections, and Coordinate-Aware Soft Hausdorff Loss using coordinate-appropriate distance metrics. Evaluated on the LISA25 challenge dataset, our ensemble achieved competitive performance with Dice coefficients of 0.72±0.17 for hippocampus and 0.85±0.05 for basal ganglia segmentation. While coordinate transformations provide principled geometric handling, inverse transformation artifacts limited their individual effectiveness. The novel loss functions demonstrate clear benefits for medical image segmentation, advancing automated analysis capabilities for portable brain MRI systems in resource-constrained environments. Source code is available at https://github.com/mahbodez/LISA25-public.

Keywords: Brain MRI segmentation · Low-field MRI · Coordinate transformations · Deep learning · Pediatric neuroimaging · LISA25 challenge

1 Introduction

Magnetic resonance imaging (MRI) is the reference modality for pediatric neuroimaging due to its excellent soft-tissue contrast and absence of ionizing radiation. Conventional 1.5T–3T scanners, however, are immobile, expensive, and require specialized shielded rooms as well as cryogenic infrastructure. Recent advances in permanent-magnet technology have enabled portable ultra-low-field (uLF) systems operating at 0.064T—notably the Hyperfine *Swoop*® device[1] can

---

[1] https://hyperfine.io/.

© The Author(s) 2026
N. Lepore and M. G. Linguraru (Eds.): LISA 2025, LNCS 16411, pp. 74–85, 2026.
https://doi.org/10.1007/978-3-032-14417-1_7

acquire point-of-care brain scans in ordinary clinical settings. Early clinical studies report that uLF images are sufficient to detect acute hemorrhage and mass effect [3], suggesting a path toward democratizing advanced neurodiagnostics in resource-constrained environments.

uLF MRI nevertheless suffers from intrinsic limitations: low signal-to-noise ratio (SNR), coarse resolution, and characteristic artifacts (e.g., pronounced B_0 inhomogeneity). These factors pose significant challenges for automated analysis pipelines trained on high-field data. The *Low-Field Pediatric Brain Image Segmentation and Quality Assurance* (LISA) challenge series [2] was established to benchmark algorithms that address these difficulties. The latest iteration, LISA25, provides fully annotated T_2-weighted ultra-low-field (uLF) volumes with two tracks: (i) artifact quality assessment and (ii) segmentation of bilateral hippocampi, caudate nuclei, lentiform nuclei, and lateral ventricles, though the lateral ventricles are not part of the primary segmentation task.

The targeted structures are clinically important for diagnosing developmental disorders [10], epilepsy [12], and hydrocephalus [11], yet their small size and low contrast make them difficult to delineate, especially at uLF resolution. Standard Cartesian convolutional networks must learn invariance to subject-specific orientation, anisotropic voxels, and shape variability, often relying on heavy data augmentation.

We advocate an alternative, *coordinate-aware* strategy that handles global geometry analytically. Each volume is first mapped to a canonical spherical or ellipsoidal lattice, effectively factoring out translation, rotation, and scale. An ensemble of SwinUNETR models is trained in Cartesian, spherical, and ellipsoidal spaces, and predictions are subsequently projected back to Cartesian voxels.

2   Methods

Objective. Let $\mathbf{X} \in \mathbb{R}^{H \times W \times D}$ denote a 3D T_2-weighted MRI volume and $\mathbf{Y} \in \{0, 1, \ldots, K\}^{H \times W \times D}$ the corresponding ground-truth segmentation map with $K+1$ semantic classes. Our objective is to learn a mapping $f : \mathbb{R}^{H \times W \times D} \to \{0, 1, \ldots, K\}^{H \times W \times D}$ that accurately delineates bilateral hippocampi, caudate nuclei, and lentiform nuclei in low-field pediatric brain MRI.

2.1   Dataset

Data Acquisition. The dataset comprises 79 low-field (0.064T) T_2-weighted MRI volumes of pediatric subjects provided by the LISA25 challenge [2]. Each volume $\mathbf{X}_i \in \mathbb{R}^{H_i \times W_i \times D_i}$ has subject-specific dimensions with typical voxel spacing of $1.0 \times 1.0 \times 1.0$ mm^3.

Ground Truth Annotations. Manual segmentations $\mathbf{Y}_i \in \{0, 1, \ldots, 8\}^{H_i \times W_i \times D_i}$ are provided for bilateral hippocampi (labels 1/2), lateral ventricles (labels 3/4), caudate nuclei (labels 5/6), and lentiform nuclei (labels 7/8), plus background (label 0). The primary task focuses on hippocampi and basal ganglia segmentation. Two sets of annotations were provided: one segmented directly from low-field images and another from high-field MRI scans subsequently registered to low-field space. The reference ground truth used for evaluation consists of the high-field-derived annotations registered to low-field images. Our training exclusively utilized these reference annotations; the low-field direct annotations were not used.

Data Split. The dataset was split into 70 volumes for training and 9 volumes for internal validation. Additionally, the challenge organizers provided 12 volumes for external validation with hidden ground truth annotations for final performance assessment.

2.2   Preprocessing Pipeline

A unified preprocessing pipeline $\mathcal{P} : \mathbb{R}^{H \times W \times D} \to \mathbb{R}^{H' \times W' \times D'}$ normalizes both intensity and spatial properties through bias-field correction [7], denoising and enhancement [6], brain extraction [8], intensity normalization, and coordinate transformation to either spherical $\mathcal{T}_{\mathrm{sph}}$ or ellipsoidal $\mathcal{T}_{\mathrm{ell}}$ domains.

2.3   Coordinate Transforms

Let $\Omega \subset \mathbb{R}^3$ denote the domain of a discrete volume $V : \Omega \cap \mathbb{Z}^3 \to \mathbb{R}$. We define two bijective transformations $\mathcal{T}_{\mathrm{sph}} : \Omega \to \mathcal{S}$ and $\mathcal{T}_{\mathrm{ell}} : \Omega \to \mathcal{E}$ that map Cartesian coordinates to regular parametric lattices, where $\mathcal{S}$ and $\mathcal{E}$ represent spherical and ellipsoidal coordinate domains respectively (Fig. 1).

Spherical Transform. The spherical transformation $\mathcal{T}_{\mathrm{sph}} : \mathbb{R}^3 \to [0, r_{\max}] \times [0, 2\pi) \times [0, \pi]$ is defined by:

$$\mathcal{T}_{\mathrm{sph}}(\mathbf{x}) = (r, \varphi, \theta), \text{ where} \tag{1}$$

$$r = \|\mathbf{x} - \mathbf{c}\|_2, \tag{2}$$

$$\varphi = \mathrm{atan2}(y - c_y, x - c_x), \tag{3}$$

$$\theta = \arccos\left(\frac{z - c_z}{\|\mathbf{x} - \mathbf{c}\|_2}\right), \tag{4}$$

for $\mathbf{x} = (x, y, z)^T \in \mathbb{R}^3$ and center $\mathbf{c} = (c_x, c_y, c_z)^T$. The inverse mapping $\mathcal{T}_{\mathrm{sph}}^{-1}$ is given by:

$$\mathcal{T}_{\mathrm{sph}}^{-1}(r, \varphi, \theta) = \mathbf{c} + r \begin{pmatrix} \sin\theta \cos\varphi \\ \sin\theta \sin\varphi \\ \cos\theta \end{pmatrix}. \tag{5}$$

Ellipsoidal Transform. The ellipsoidal transformation $\mathcal{T}_{\text{ell}} : \mathbb{R}^3 \to \mathbb{R}^3$ first fits an ellipsoid to the brain mask via principal component analysis (PCA) to obtain center $\mathbf{c}$, rotation matrix $\mathbf{R}$, and scaling matrix $\mathbf{S}$. Each point is then normalized to unit ellipsoid coordinates:

$$\tilde{\mathbf{x}} = \mathbf{S}\mathbf{R}^T(\mathbf{x} - \mathbf{c}), \tag{6}$$

followed by spherical coordinate computation from $\tilde{\mathbf{x}}$. The inverse mapping is:

$$\mathcal{T}_{\text{ell}}^{-1}(r, \varphi, \theta) = \mathbf{c} + \mathbf{R}\mathbf{S}^{-1}\mathcal{T}_{\text{sph}}^{-1}(r, \varphi, \theta). \tag{7}$$

Figure 2 demonstrates the visual effects of these coordinate transformations on actual pediatric brain MRI data.

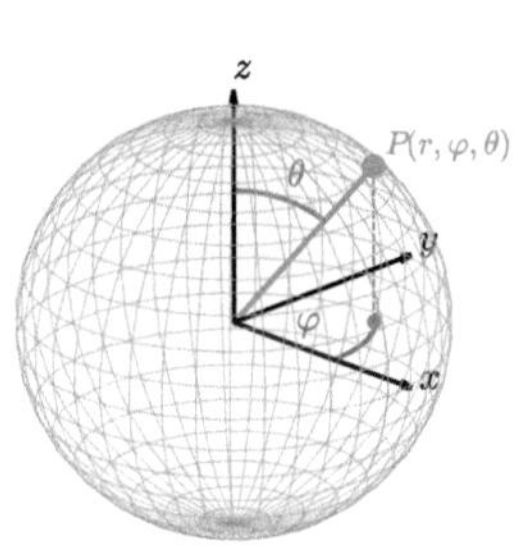

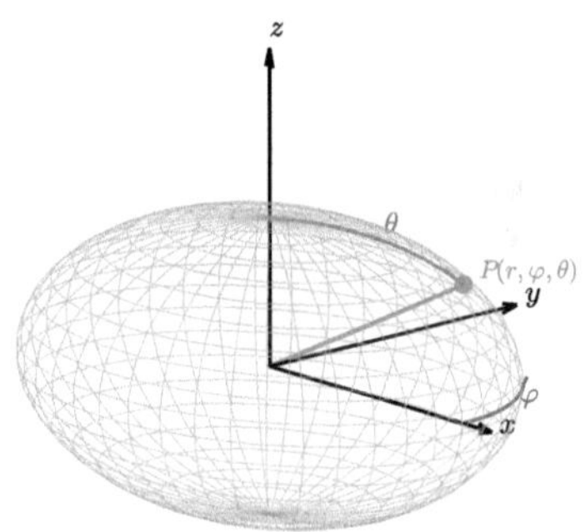

(a) Spherical coordinate system showing the relationship between Cartesian coordinates (x, y, z) and spherical coordinates (r, φ, θ), where φ is the longitude and θ is the latitude.

(b) Ellipsoidal coordinate transformation normalizing subject-specific geometry by mapping points to a standardized ellipsoidal space before applying spherical coordinates.

Fig. 1. Coordinate transformation systems used in our approach. Both transformations demonstrate the geometric mappings that enable factoring out translation, rotation, and scale variations.

2.4 Data Augmentation

Augmentations are applied in the transformed coordinate space to preserve geometric properties, including hemispheric flip (reflection $\varphi \mapsto \pi - \varphi$ with bilateral label swapping), longitudinal rotation (cyclic shift $\varphi \mapsto (\varphi + \delta) \bmod 2\pi$), radial scaling ($r \mapsto \gamma r$), gamma correction ($\mathbf{V} \mapsto \mathbf{V}^\gamma$), and robust rescaling using 5th and 95th percentiles.

2.5   Model Architecture

We employ a Swin Transformer U-Net (SwinUNETR) [4] implemented in MONAI [5] using the PyTorch framework, combining hierarchical feature learning with U-Net skip connections. The network uses 48-dimensional features with [3, 6, 12, 24] attention heads across four encoder stages, instance normalization, and outputs 9 classes (background + 8 anatomical structures) as logits $\in \mathbb{R}^{9 \times H \times W \times D}$.

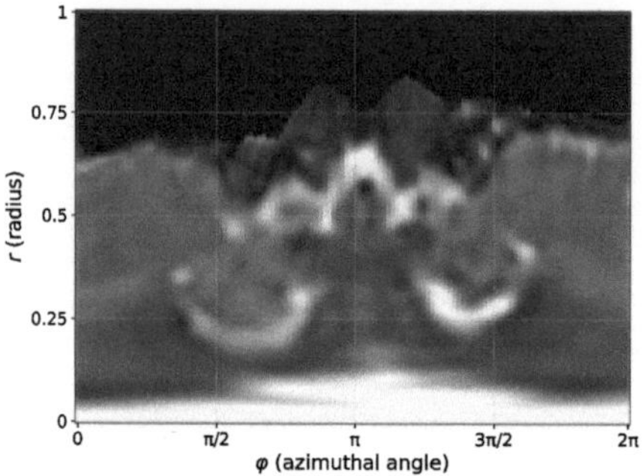

(a) Spherical coordinate-transformed brain volume showing the middle axial slice from a pediatric T_2-weighted uLF MRI (patient 0015).

(b) Ellipsoidal coordinate-transformed brain volume showing the corresponding middle axial slice from the same patient, demonstrating the geometric normalization effect.

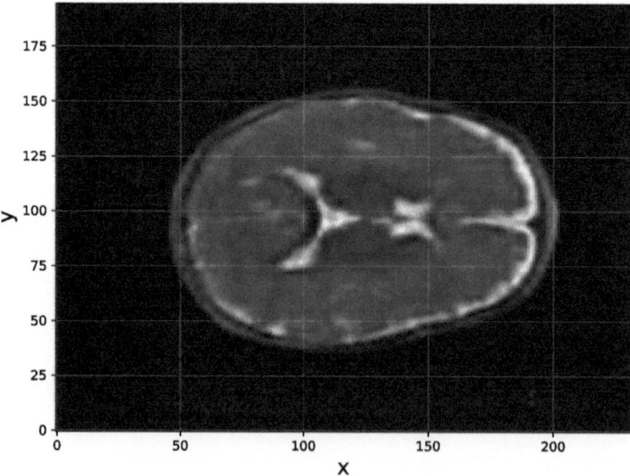

(c) Cartesian (axial) middle slice of the same patient for reference.

Fig. 2. Three middle slices with the same θ index from pediatric brain MRI volumes: spherical, ellipsoidal, and the corresponding Cartesian (axial) reference. The coordinate transforms alter the appearance of anatomical structures while factoring out global geometry.

2.6   Loss Function and Optimization

Composite Loss Function. The training objective combines multiple complementary loss terms with epoch-dependent weights:

$$\mathcal{L}(\theta, t) = \sum_k \alpha_k(t) \cdot \mathcal{L}_k(\theta) \tag{8}$$

where k includes Cross-Entropy, Generalized Dice, Projection Dice, and coordinate-specific Soft Hausdorff terms, with $\alpha_k(t)$ representing time-varying weights for each component.

Projection Dice Loss. Traditional 3D Dice loss operates on the full volumetric segmentation, which can be sensitive to small misalignments and may not adequately capture the global shape characteristics crucial for anatomical structure segmentation. We introduce a *Projection Dice Loss* that evaluates segmentation quality through 2D orthogonal projections, providing complementary shape-aware supervision.

Methodology. For each class c (optionally excluding background), we form 2D "silhouette" projections along the three orthogonal axes using a differentiable soft-OR reduction. Given per-voxel class probabilities $p \in [0, 1]$, the soft projection along dimension d is computed as:

$$\mathrm{proj}_d = 1 - \prod_{i \in \mathrm{axis}_d} (1 - p_i) \tag{9}$$

This operation is implemented stably in log-space using:

$$\mathrm{proj}_d = 1 - \exp\left(\sum_{i \in \mathrm{axis}_d} \log(1 - p_i) \right) \tag{10}$$

For ground-truth labels, we use hard projections via maximum operations: $\mathrm{proj}_d^{gt} = \max_{i \in \mathrm{axis}_d} g_i$.

Loss Computation. We compute standard Dice coefficients between predicted and ground-truth projections for each of the three orthogonal axes:

$$\mathrm{Dice}_d = \frac{2 \sum \mathrm{proj}_d \cdot \mathrm{proj}_d^{gt} + \epsilon}{\sum \mathrm{proj}_d + \sum \mathrm{proj}_d^{gt} + \epsilon} \tag{11}$$

The final Projection Dice loss averages across all three axes:

$$\mathcal{L}_{\mathrm{proj}} = 1 - \frac{1}{3} \sum_{d \in \{x,y,z\}} \mathrm{Dice}_d \tag{12}$$

Coordinate-Aware Soft Hausdorff Loss. Classical Hausdorff distance (HD) measures the largest point-to-set error in *Euclidean* space. Directly applying it to data that have been mapped to spherical or other curvilinear grids conflates radial and angular errors and wastes memory on interior voxels. We therefore formulate two *coordinate-specific Soft Hausdorff* losses that (i) operate only on surface points, (ii) use the natural metric of the parameterization, and (iii) remain fully differentiable.

Soft-extreme Framework. Given the predicted surface points $\mathcal{A} = \{a_i\}_{i=1}^{M}$ (with per-point confidences $w_i \in (0,1]$) and the ground-truth surface points $\mathcal{B} = \{b_j\}_{j=1}^{K}$, let $d(\cdot,\cdot)$ be a coordinate-appropriate distance. We replace the hard $\max - \min$ of HD by a *log-sum-exp* (LSE) smoothing:

$$\tilde{d}_i^{\min} = -\tfrac{1}{\beta} \log \sum_{j=1}^{K} \exp\!\big[-\beta\, d(a_i, b_j)\big], \tag{13}$$

$$\tilde{h}_{\alpha,\beta}(\mathcal{A} \to \mathcal{B}) = \tfrac{1}{\alpha} \log \sum_{i=1}^{M} w_i \, \exp\!\big[\alpha\, \tilde{d}_i^{\min}\big], \tag{14}$$

where $\beta > 0$ (inner soft-minimum) and $\alpha > 0$ (outer soft-maximum) control the sharpness. Setting $\alpha, \beta \to \infty$ recovers the exact HD, while $\alpha, \beta \approx 5$ yields a Chamfer-like mean surface distance.

The symmetric soft HD we minimize is

$$\mathcal{L}_{\mathrm{HD}} = \frac{1}{2}\Big[\tilde{h}_{\alpha,\beta}(\mathcal{A} \to \mathcal{B}) + \tilde{h}_{\alpha,\beta}(\mathcal{B} \to \mathcal{A})\Big]. \tag{15}$$

Cartesian Metric. For voxel grids in (x, y, z) the standard Euclidean metric is retained,

$$d_{\mathrm{cart}}(\mathbf{a}, \mathbf{b}) = \|\mathbf{a} - \mathbf{b}\|_2 + \varepsilon, \tag{16}$$

with $\varepsilon = 10^{-6}$ to prevent zero gradients when $\mathbf{a} = \mathbf{b}$.

Spherical Metric. For spherical/ellipsoidal tuples (r, φ, θ) we couple the radial difference with the great-circle angle:

$$d_{\mathrm{sph}}(a, b) = \sqrt{(r_a - r_b)^2 + (\lambda\, \bar{r}\, \sigma)^2 + \varepsilon}, \tag{17}$$

with $\bar{r} = \tfrac{1}{2}(r_a + r_b)$, λ a unit-free trade-off hyperparameter ($\lambda = 1$ in all experiments), and

$$\sigma = \arccos\!\big(\cos\theta_a \cos\theta_b + \sin\theta_a \sin\theta_b \cos(\varphi_a - \varphi_b)\big). \tag{18}$$

Equation (17) is the *exact* Euclidean distance expressed in spherical coordinates, so the loss still converges to true HD for $\alpha, \beta \to \infty$.

Surface Extraction and Sampling. We first isolate *boundary voxels* by a 26-connected morphological test: a foreground voxel is marked as surface if the $3 \times 3 \times 3$ binary sum of its neighborhood is < 27. From each class we keep at most $N_{\text{pred}} = 4096$ predicted surface voxels, chosen by the top-k confidence scores after a liberal threshold $\tau = 10^{-3}$, and $N_{\text{gt}} = 4096$ uniformly-sampled ground-truth surface voxels. This limits the pairwise matrix in (13) to $k^2 \leq 16M$ entries without altering the HD value, which depends only on the *worst* points.

Confidence Weighting. Prediction confidences enter (14) as $w_i = P(\text{class}|a_i)$. Algebraically this is equivalent to adding $-\frac{1}{\beta} \log w_i$ inside the soft-minimum, so highly confident yet misplaced voxels receive the largest gradients, while low-confidence outliers are softly down-weighted.

Optimization. We minimize $\mathcal{L}(\theta)$ using AdamW [14] with learning rate $\eta = 10^{-4}$ and weight decay $\lambda = 10^{-4}$. The learning rate schedule consists of linear warm-up, followed by a steady learning rate, and then cosine annealing. Mixed-precision training using `bfloat16` and gradient accumulation is performed.

2.7   Training and Inference

Training. Models are trained for 100 epochs with effective batch size 32 (4 per graphics processing unit (GPU) $\times$ 8 accumulation), AdamW optimizer, and mixed-precision on dual RTX 4090 GPUs ($\sim$8 h per model).

Ensemble Inference. Let f_{cart}, f_{sph}, and f_{ell} denote the trained models in Cartesian, spherical, and ellipsoidal coordinates, respectively. For a test volume $\mathbf{X}$, we compute three predictions:

$$\hat{\mathbf{Y}}_{\text{cart}} = \arg \max_k (f_{\text{cart}}(\mathbf{X})), \tag{19}$$

$$\hat{\mathbf{Y}}_{\text{sph}} = \arg \max_k (\mathcal{T}_{\text{sph}}^{-1}(f_{\text{sph}}(\mathcal{T}_{\text{sph}}(\mathbf{X})))), \tag{20}$$

$$\hat{\mathbf{Y}}_{\text{ell}} = \arg \max_k (\mathcal{T}_{\text{ell}}^{-1}(f_{\text{ell}}(\mathcal{T}_{\text{ell}}(\mathbf{X})))), \tag{21}$$

where $\mathcal{T}_{\text{sph}}$ and $\mathcal{T}_{\text{ell}}$ correspond to forward spherical and ellipsoidal coordinate transformations, while $\mathcal{T}_{\text{sph}}^{-1}$ and $\mathcal{T}_{\text{ell}}^{-1}$ denote their inverse projections.

The final prediction uses majority voting with tie-breaking. Here, a tie occurs when no single segmentation label receives 2 or more votes; in that case, we revert to the Cartesian model's prediction:

$$\hat{y}(\mathbf{x}) = \begin{cases} \hat{\mathbf{Y}}_{\text{cart}}(\mathbf{x}) & \text{if no label gets 2 or more votes,} \\ \text{mode}\{\hat{\mathbf{Y}}_{\text{cart}}(\mathbf{x}), \hat{\mathbf{Y}}_{\text{sph}}(\mathbf{x}), \hat{\mathbf{Y}}_{\text{ell}}(\mathbf{x})\} & \text{otherwise.} \end{cases} \tag{22}$$

Inference employs sliding-window prediction with configurable overlap to handle memory constraints and boundary effects. Each model contains approximately 73 million parameters.

2.8   Evaluation Metrics

Performance assessment employs standard segmentation metrics: Dice similarity coefficient (DSC), Hausdorff distance (HD), 95th percentile Hausdorff distance (HD95), average symmetric surface distance (ASSD), and relative volume error (RVE).

3   Results

Our coordinate transform-based ensemble approach demonstrates competitive performance on both hippocampus and basal ganglia segmentation tasks. Results are reported on the 12 external validation volumes provided by the challenge organizers. For hippocampus segmentation (Task 2a), the ensemble model achieved a mean Dice coefficient of 0.72 ± 0.17, outperforming individual models including Cartesian (0.71 ± 0.19), spherical (0.70 ± 0.12), and ellipsoidal (0.65 ± 0.16) variants (Table 1). The ensemble also showed improvements in surface-based metrics, with Hausdorff distance of 5.40 ± 8.01 mm and 95th percentile Hausdorff distance of 1.87 ± 1.13 mm.

For basal ganglia segmentation (Task 2b), all models performed substantially better, with the ensemble achieving a Dice coefficient of 0.85 ± 0.05 (Table 1). Interestingly, the Cartesian model matched the ensemble performance (0.85 ± 0.05) for this task, while spherical (0.83 ± 0.06) and ellipsoidal (0.80 ± 0.05) models showed slightly lower but still competitive results. Surface distances were consistently lower for basal ganglia, with ensemble ASSD of 0.51 ± 0.26 mm compared to 0.66 ± 0.60 mm for hippocampus.

Table 1. Performance of models on hippocampus (Task 2a) and basal ganglia (Task 2b) segmentation evaluated on 12 external validation volumes. Values are mean±SD; ↑ means higher is better and ↓ lower is better.

Task	Model	DSC ↑	HD ↓	95HD ↓	ASSD ↓	RVE ↓
Task 2a	Ensemble	**0.72** ± 0.17	**5.40** ± 8.01	**1.87** ± 1.13	**0.66** ± 0.60	0.18 ± 0.11
	Cartesian	0.71 ± 0.19	5.63 ± 8.10	2.06 ± 1.36	0.74 ± 0.82	**0.13** ± 0.08
	Spherical	0.70 ± 0.12	5.58 ± 8.02	1.97 ± 0.79	**0.66** ± 0.39	0.20 ± 0.15
	Ellipsoidal	0.65 ± 0.16	6.33 ± 7.80	2.58 ± 1.21	0.84 ± 0.54	0.27 ± 0.18
Task 2b	Ensemble	**0.85**±0.05	3.35 ± 1.18	1.89 ± 1.09	**0.51** ± 0.26	0.17±0.08
	Cartesian	**0.85**±0.05	**3.24** ± 0.99	**1.79** ± 0.89	**0.51** ± 0.23	**0.15** ± 0.07
	Spherical	0.83 ± 0.06	3.70 ± 1.28	2.02 ± 1.13	0.55 ± 0.30	0.17 ± 0.09
	Ellipsoidal	0.80 ± 0.05	5.48 ± 3.63	2.66 ± 1.30	0.74 ± 0.24	0.27 ± 0.12

4   Discussion

This work introduces a coordinate transform-based approach for low-field pediatric brain MRI segmentation with three key contributions.

Coordinate Transform Strategy. By mapping brain volumes to spherical and ellipsoidal coordinate systems, we analytically factor out translation, rotation, and scale variations that typically require extensive data augmentation. This is particularly valuable for pediatric neuroimaging where anatomical variability is substantial and training data is limited.

Novel Loss Functions. The Projection Dice Loss provides shape-aware supervision through 2D orthogonal projections, capturing global structure characteristics beneficial for small structures like the hippocampus. The Coordinate-Aware Soft Hausdorff Loss operates on surface points using coordinate-appropriate metrics, avoiding conflation of radial and angular errors when applying Euclidean distances to transformed spaces.

Performance and Limitations. The ensemble achieved competitive performance (hippocampus DSC: 0.72 ± 0.17; basal ganglia DSC: 0.85 ± 0.05). However, spherical and ellipsoidal models underperformed relative to Cartesian, likely due to interpolation artifacts introduced during inverse transformation back to Cartesian space. These artifacts are more pronounced for ellipsoidal transforms due to the complex transformation pathway. Interestingly, the coordinate transformations showed more promise for hippocampal segmentation, where individual transformed models (spherical: 0.70 ± 0.12, ellipsoidal: 0.65 ± 0.16) achieved relatively closer performance to Cartesian (0.71 ± 0.19) compared to basal ganglia structures. This difference may be attributed to the anatomical characteristics: the hippocampus has an inherently curved, elongated structure that naturally aligns with spherical and ellipsoidal coordinate systems, potentially making its geometric features more apparent after transformation. In contrast, basal ganglia structures (caudate and lentiform nuclei) have more compact, roughly spherical morphologies that may not benefit as substantially from coordinate transformation, explaining why the Cartesian model matched ensemble performance (0.85 ± 0.05) for this task.

5   Conclusion

We presented a coordinate transform-based approach combining spherical/ellipsoidal transformations with novel Projection Dice and Coordinate-Aware Soft Hausdorff losses for low-field pediatric brain MRI segmentation. The ensemble achieved competitive performance (hippocampus DSC: 0.72 ± 0.17; basal ganglia DSC: 0.85 ± 0.05), though inverse transformation artifacts limited the effectiveness of coordinate transforms. The novel loss functions provide clear benefits for shape-aware supervision and surface-based optimization. This work advances automated analysis of portable brain MRI systems, supporting point-of-care neuroimaging in resource-limited settings.

Disclosure of Interests. The author declares that there is no conflict of interest.

References

1. LISA25 Challenge Homepage, https://lisa-challenge.github.io/, Accessed 12 July 2025
2. LISA challenge organizers: LISA25 - low-field pediatric brain image segmentation and quality assurance challenge. In: MICCAI 2025. LNCS, pp. 1–13. Springer, Heidelberg (2025)
3. Sheth, K.N., Mazurek, M.H., Yuen, M.M., et al.: Assessment of brain injury using portable, low-field magnetic resonance imaging at the bedside of critically ill patients. JAMA Neurol. **78**(1), 41–47 (2021)
4. Hatamizadeh, A., Nath, V., Tang, Y., et al.: Swin UNETR: swin transformers for semantic segmentation of brain tumors in MRI images. In: Crimi, A., Bakas, S. (eds.) BrainLes 2022, LNCS, vol. 13039, pp. 272–284. Springer, Cham (2022). https://doi.org/10.1007/978-3-031-06039-7_23
5. MONAI Consortium: MONAI – Medical Open Network for AI, https://monai.io/, Accessed 12 July 2025
6. Maggioni, M., Katkovnik, V., Egiazarian, K., Foi, A.: Nonlocal transform-domain filter for volumetric data denoising and reconstruction. IEEE Trans. Image Process. **22**(1), 119–133 (2013)
7. Tustison, N.J., Avants, B.B., Cook, P.A., et al.: N4ITK: improved N3 bias correction. IEEE Trans. Med. Imaging **29**(6), 1310–1320 (2010)
8. Smith, S.M.: Fast robust automated brain extraction. Hum. Brain Mapp. **17**(3), 143–155 (2002)
9. Zhou, W., Li, J., Wang, X., et al.: Infant hippocampal segmentation in ultra-low-field mri using external datasets with diverse field strengths. In: Low Field Pediatric Brain MRI Segmentation and QA (LISA). LNCS 15515, pp. 28–37. Springer, Cham (2025)
10. Qiu, T., Chang, C., Li, Y., et al.: Two years changes in the development of caudate nucleus are involved in restricted repetitive behaviors in 2- to 5-year-old children with autism spectrum disorder. Dev. Cogn. Neurosci. **19**, 137–143 (2016)
11. Gruber, M.D., Unadkat, P., Morales, D.M., et al.: Ultra-low-field portable MRI for assessing ventricular size in pediatric hydrocephalus: a feasibility study. J. Neurosurg. Pediatr. **36**(1), 11–19 (2025)
12. Mansour, A.R., Shetty, T., Chandra, V., et al.: Validation of automatic MRI hippocampal subfield segmentation by histopathology in epilepsy surgery patients. Epilepsia **62**(8), 1904–1916 (2021)
13. Sudre, C.H., Li, W., Vercauteren, T., Ourselin, S., Jorge Cardoso, M.: Generalised dice overlap as a deep learning loss function for highly unbalanced segmentations. In: Deep Learning in Medical Image Analysis and Multimodal Learning for Clinical Decision Support, pp. 240–248. Springer (2017)
14. Loshchilov, I., Hutter, F.: Decoupled weight decay regularization. In: International Conference on Learning Representations (ICLR) (2019)

Atlas-Augmented Semantic Segmentation for Robust Ultra-Low-Field Pediatric Brain Imaging

Kostiantyn Lavronenko[1,2(✉)] , Rueveyda Yilmaz[1] , Zhu Chen[1] ,
Johannes Stegmaier[1] , and Volkmar Schulz[1,2]

[1] RWTH Aachen University, Aachen, Germany
[2] Fraunhofer MEVIS, Bremen, Germany
`kostiantyn.lavronenko@mevis.fraunhofer.de`

Abstract. Low-field MRI offers a portable, cost-effective alternative to conventional high-field scanners but suffers from reduced signal-to-noise ratio and spatial inhomogeneity, which compromise the accuracy and consistency of automated brain structure segmentation. In this work, we introduce atlas-augmented deep learning models that integrate probabilistic anatomical priors to enhance the delineation of pediatric hippocampus and basal ganglia in ultra-low-field MRI (0.064 T). We evaluate seven pipelines on the LISA 2025 dataset (79 T2-weighted scans): baseline VNet, nnU-Net, and MedSAM2 variants (2D and 3D decoders), as well as atlas-augmented VNet, atlas-augmented nnU-Net, and atlas-augmented MedSAM2-3D. For VNet and MedSAM2-3D, probabilistic maps from the Pauli and Harvard-Oxford atlases are encoded and fused with intermediate feature maps, while nnU-Net ingests priors as additional input channels. Baseline nnU-Net attains mean DSCs of 0.71 for hippocampus and 0.86 for basal ganglia; atlas augmentation yields modest hippocampal gains (HD95 ↓0.05, ASSD ↓0.06) and more pronounced improvements in basal ganglia segmentation, reflecting richer prior information for larger structures. VNet and MedSAM2 variants exhibit limited hippocampal benefit, highlighting the strength of nnU-Net's adaptive framework. Our findings establish atlas-augmented nnU-Net as a new benchmark for robust segmentation in resource-constrained, low-field imaging environments. The code for our methods will be publicly accessible after the successful publication of the paper here: https:// github.com/mackostya/deepatlas-ulf-seg.

Keywords: 3D Segmentation · Low-field MRI · Brain Imaging

1 Introduction

Low-field magnetic resonance imaging (MRI) has recently gained traction as a cost-effective and portable alternative to conventional high-field systems [1]. By operating at lower field strengths (e.g., 0.064 T), these scanners offer reduced

© The Author(s) 2026
N. Lepore and M. G. Linguraru (Eds.): LISA 2025, LNCS 16411, pp. 86–97, 2026.
https://doi.org/10.1007/978-3-032-14417-1_8

acquisition and maintenance costs [2], improved safety profiles, and enhanced accessibility—particularly in resource-limited settings or pediatric applications where sedation may be undesirable [3,4]. Nevertheless, the intrinsic physical constraints of low-field imaging give rise to reduced signal-to-noise ratio (SNR), increased magnetic field inhomogeneity, and longer scan durations [5]. Consequently, the resulting image quality falls short of that achieved with standard high-field MRI, posing substantial challenges for accurate description of brain structures in ultra-low-field (uLF) pediatric data. Furthermore, pediatric brain MRI images pose an additional challenge for segmentation due to the lower contrasts in gray and white matter in comparison to the older cohort [6]. To this end, in LISA 2024 challenge, hippocampus segmentation in ultra-low-field MRI was approached through a variety of strategies. Most teams relied on U-Net-based frameworks such as nnU-Net [7,8] with additional modification techniques such as adaptation of the data to high-field, training with a mixture of different datasets, pseudo-labelling, etc. By using an atlas-based segmentation approach, Sundaresan et al. achieved a better performance compared to a 3D UNet-based segmentation method likely due to the low contrast in the images [9]. Alternatively, Peiris et al. proposes a dual training pipeline by combining features from the original and high-frequency-filtered images [10].

In this study, we introduce an atlas-augmented segmentation pipeline for uLF MRI that enhances three backbone architectures—VNet [11], MedSAM2 [12], and nnU-Net [13] with spatial priors from Montreal Neurological Institute (MNI) brain space registration [14]. Each subject's T2 volume is nonrigidly registered to the MNI pediatric template using Advanced Normalization Tools (ANTS) [15], producing probabilistic maps for bilateral hippocampi and basal ganglia. Our experiments show that atlas augmentation consistently enhances segmentation accuracy for both hippocampus and basal ganglia. Notably, the atlas-augmented nnU-Net achieves the highest performance across all metrics, establishing a new state-of-the-art on the LISA 2025 Task 2 dataset. These findings highlight the value of combining data-driven architectures with anatomical priors for robust uLF pediatric MRI segmentation.

2   Data

The LISA 2025 dataset comprises 79 volumetric T2-weighted MRI scans of pediatric brains, acquired on a Hyperfine 0.064 T [16] low-field system. Each volume is accompanied by binary masks for nine anatomical structures—most notably the bilateral hippocampi and basal ganglia. These masks were manually delineated by expert neuroimagers on the high field volumes and coregistered through a 9-point linear registration to the low-field domain to ensure annotation fidelity. Due to the uLF acquisition, the images exhibit reduced signal-to-noise ratio and pronounced magnetic field inhomogeneities relative to conventional high-field data, presenting a challenge for segmentation algorithms. All scans and corresponding masks are provided in NIfTI (.nii.gz) format.

An example subject's low-field T2 image and corresponding ground truth segmentation masks are shown in Fig. 1. Each mask encodes eight labels: left and

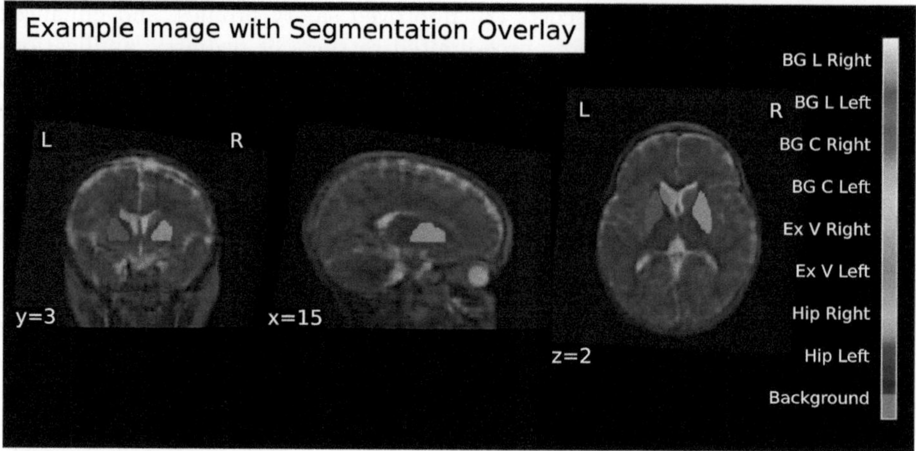

Fig. 1. An example volume with LISA 25 Challenge ground truth segmentation overlays. Labels: BG L (basal ganglia lentiform), BG C (basal ganglia caudate), Ex V (extra ventricle), Hip (hippocampus).

right hippocampus (Subtask 2a), left and right lentiform nucleus and caudate nucleus (Subtask 2b), and left and right lateral ventricles (auxiliary structures). All described masks were used for the training of the models. Only the labels registered from the high field were used as ground truth in our work. All volumes and masks share identical spatial resolution of $1\,mm^3$. The cohort of 79 scans was divided into 90% for training (71 cases) and 10% for internal validation (8 cases). The training subset is used to optimize model parameters, while the validation subset provides intermediate performance assessment and checkpoint selection during training. It is important to note that nnU-Net performs its own internal cross-validation split and preprocessing pipeline, independently determining its training and validation folds.

The official validation set of 12 subjects is used for final evaluation, with results submitted to the LISA 2025 challenge platform. The evaluation for each subtask concerned only the labels, that correspond to the subtask, meaning that the ventricle labels were not used for the final evaluation.

3   Model Selection Motivation

For model selection, insights from the LISA 2024 Challenge results [17] were taken into account. In that context, the nnU-Net model [13] was successfully applied to pediatric image segmentation tasks [7,8,18], which motivated its consideration for integration in our work. Additionally, none of the participating teams employed foundational models, which are increasingly gaining attention in the research community. This motivated us to explore the recent foundational segmentation model MedSAM2 [12] for the low-field segmentation problem. Finally, the VNet model was included as a baseline for comparison, given its effective dual-view learning architecture [10].

Our evaluation includes seven segmentation pipelines on the LISA 2025 Task 2 dataset: (1) the baseline VNet [11] and its atlas-augmented counterpart; (2) the baseline nnU-Net [13] and its atlas-augmented variant; and (3) three MedSAM2-based models that share the same pretrained image encoder but differ in decoding: a 2D decoder, a 3D decoder, and an atlas-augmented 3D decoder. Atlas priors are incorporated only into the 3D decoder variant. Detailed descriptions of each architecture and their training configurations are provided in Chap. 4, with the overall methodology depicted in Fig. 2.

4   Methods

All models were trained using a composite loss that combines voxel-wise Cross-Entropy (CE) and the Dice loss, a formulation well suited to address class imbalance in segmentation tasks. The total loss is defined as

$$\mathcal{L}_{\text{total}} = \mathcal{L}_{\text{CE}} + \mathcal{L}_{\text{Dice}}, \tag{1}$$

where

$$\mathcal{L}_{\text{CE}} = -\frac{1}{N} \sum_{i=1}^{N} \sum_{c=1}^{C} g_c(v_i) \log p_c(v_i),$$

and the Dice loss is given by

$$\mathcal{L}_{\text{Dice}} = 1 - \frac{2 \sum_{i=1}^{N} \sum_{c=1}^{C} p_c(v_i)\, g_c(v_i)}{\sum_{i=1}^{N} \sum_{c=1}^{C} p_c(v_i) + \sum_{i=1}^{N} \sum_{c=1}^{C} g_c(v_i)}, \tag{2}$$

where, N is the total number of voxels, C is the number of classes, $p_c(v_i)$ denotes the predicted probability for voxel v_i belonging to class c and $g_c(v_i)$ is the one-hot encoded ground-truth label. We adopt the Dice loss implementation by MONAI [19].

For VNet and MedSAM2 models, all input volumes were resampled to a uniform grid of 128^3 voxels, matching the native input size of both architectures. Data augmentation—including random elastic deformations and intensity perturbations was applied via TorchIO [20] to improve generalization. MedSAM2-based models processed slices in SAR (Superior Anterior Right) orientation to align with the pretrained encoder; outputs were subsequently reoriented to RAS space. Training was conducted on an NVIDIA Ada 5000 GPU (24 GB VRAM) using the Adam optimizer (learning rate 1×10^{-3}), a batch size varying from 1 to 8 (model-dependent), and up to 1000 epochs.

4.1   VNet

VNet [11] is a 3D convolutional encoder-decoder architecture designed for volumetric medical image segmentation. It follows the U-Net paradigm with symmetric downsampling and upsampling pathways connected via skip connections to maintain spatial detail across scales. Compared to the original U-Net, VNet incorporates residual connections within each block and a deeper network topology. In the encoder, successive layers of 3D convolutions, instance normalization, and ReLU activations reduce spatial resolution via strided convolutions; the decoder uses transposed convolutions and skip-connection concatenations to reconstruct fine-grained boundaries. We use the PyTorch implementation by Adaloglou et al. [21], which comprises approximately 45.6 million parameters.

4.2   NnU-Net

nnU-Net [13] is a self-configuring segmentation framework that automatically adapts its U-Net-based architecture, preprocessing pipeline, and training hyperparameters to a given dataset. Through nested cross-validation on the full set of 79 scans, nnU-Net identifies optimal patch sizes, network depth, normalization schemes, and data augmentation strategies, thereby eliminating manual model design. For the LISA 2025 volumes, it resamples and crops inputs to $112 \times 160 \times 128$ voxels, applies case-wise intensity normalization, and performs built-in five-fold cross-validation to configure both architecture and training. We utilize the "3d fullres" configuration for our task, which integrates 3d convolutional blocks for the volume segmentation. The autonomous splitting yields five independent training folds of the 79 volumes, four of which with 63 training and 16 validation cases and one with 64 training and 15 validation cases.

4.3   MedSAM2

MedSAM2 [12] is a promptable foundational model for medical image segmentation, pretrained on 455,000 3D image-mask pairs and 76,000 video frames. Its core architecture extends the SAM2 vision transformer backbone [22], capturing long-range spatial dependencies crucial for delineating complex anatomy. For LISA 2025 Task 2, we repurpose only the pretrained MedSAM2 image encoder to extract high-level features from each axial slice of the 128^3 resampled low-field volumes. Although its pretraining decomposed 3D data into 2D inputs, MedSAM2 retains SAM2's native 2D format; accordingly, we sequentially process slices in SAR orientation—using the first volume dimension for slicing—and then reassemble segmentation outputs in RAS space. By omitting the prompt and memory-bank modules, this uni-directional, slice-wise encoding yields efficient and robust features for subsequent decoding.

2D FPN Decoder (MS2+2D). We employ a compact FPN-style decoder on each axial MedSAM2 feature map. At each of three scales, 3×3 2D convolutions (with instance normalization and ReLU) merge via lateral connections, and bilinear interpolation restores spatial resolution. A final 1×1 convolution produces per-slice logits, which are simply stacked to reconstruct the 3D segmentation volume. This design omits transpose convolutions for efficiency and comprises approximately 27.3 million parameters.

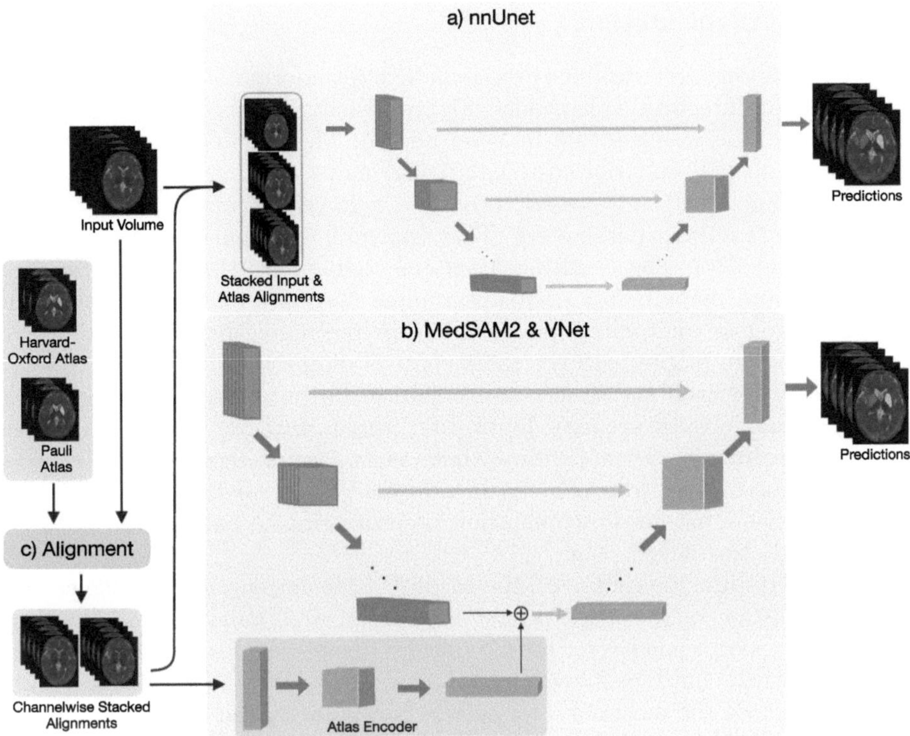

Fig. 2. Atlas-augmented segmentation architectures. (**a**) nnU-Net pipeline for segmentation. Atlas augmentation is applied at the volume input level adding the Pauli and Harvard-Oxford Atlas as additional channels. (**b**) MedSAM2's and VNet's pipeline for segmentation, including the separate Atlas Encoder for the incorporation of the priors. (**c**) Atlas registration: low-field volumes are aligned to Harvard-Oxford and Pauli probabilistic atlases. Atlas features are appended as additional input channels for nnU-Net or passed through Atlas Encoder for VNet/MedSAM2.

3D Decoder (MS2+3D). To exploit full volumetric context, we reorganize the 2D feature maps from the MedSAM2 encoder into a $C \times D \times H \times W$ tensor and process them through a 3D decoder inspired by Bui et al.'s SAM3D design [23], here applied to SAM2 features. The decoder comprises three successive blocks of 3D convolutions (kernel size 3), group normalization, and LeakyReLU activations, interleaved with trilinear upsampling stages. At each resolution, skip connections fuse low- and high-level features in an FPN-style hierarchy, facilitating precise boundary delineation. The final $1 \times 1 \times 1$ convolution produces dense volumetric logits, which after softmaxing generate the segmentation mask. This architecture contains approximately 37.8 million parameters, reflecting the increased capacity required by volumetric operations.

4.4 Atlas Augmentation

For each backbone, we introduce probabilistic atlas information in a manner that respects its architectural constraints. Figure 2 illustrates the general atlas augmentation process in our setup. In VNet and the MedSAM2 3D decoder (Fig. 2 (b)), the Pauli [24] and Harvard-Oxford [25] maps are first resampled to the input grid (Fig. 2 (c)) and passed through a lightweight convolutional encoder comprising 2.9 million parameters. This approach is an adaptation from Liu, Huabing et al. [26]. The resulting atlas embeddings are then added to intermediate feature maps before each upsampling stage, enabling the network to learn how to fuse anatomical priors with learned representations. Since both networks' feature maps share the same channel dimensionality, we use a single atlas encoder for both and simply interpolate its output along the depth axis to match each model's feature map depth. In contrast, nnU-Net (Fig. 2 (a)) ingests atlas volumes directly by appending them as additional input channels alongside the original image (from Fig. 2 (c) to Input Volumes before being passed to nnU-Net), leveraging its self-configuring preprocessing to balance the influence of image and atlas information.

Preliminary investigation with the MedSAM2 image encoder has shown that the incorporation of the atlas priors through an additional channel does not improve the relative metrics for the Challenge. Therefore, alternative atlas incorporation methods have been investigated, of which adaptation of atlas encoding as in [26] showed the most success. Furthermore, due to self-constructability of the nnU-Net the addition of the custom atlas encoding is challenging, when one wants to fully utilize the potential of the framework. Therefore, for nnU-Net atlas priors were incorporated through an additional channel, which matches the expectation of the nnU-Net network.

This strategy yields three atlas-augmented variants: VNet(A), MedSAM2(A), and nnU-Net(A).

4.5 Evaluation Metrics

Following the training, all models were assessed on the official LISA25 validation directly on the challenge platform, using the challenge metrics [27,28]: the Dice Similarity Coefficient (DSC), Hausdorff Distance (HD), 95th percentile Hausdorff Distance (HD95), Average Symmetric Surface Distance (ASSD), and Relative Volume Error (RVE).

5 Results

All models were trained end-to-end to predict nine segmentation classes: background; left and right hippocampus (labels 1–2, Subtask 2a); left and right lateral ventricles (labels 3–4, auxiliary supervision only); and left and right lentiform and caudate nuclei (labels 5–8, Subtask 2b). Overall, the baseline nnU-Net outperformed both VNet and all MedSAM2 variants across mean DSC, HD, HD95,

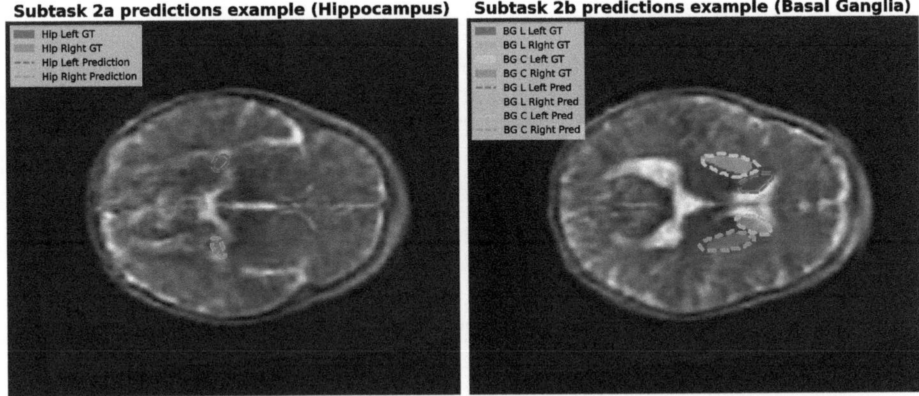

Fig. 3. Example of predictions with nnU-Net(A) model for the Subtask 2a (left) and Subtask 2b (right).

ASSD, and RVE. Incorporation of atlas priors into nnU-Net yielded further improvements, with the atlas-augmented nnU-Net achieving the highest DSC and lowest distance errors among all models. Table 1 reports the Subtask 2a results, and Table 2 reports the Subtask 2b results (Fig. 3).

5.1   Results 2a: Hippocampus

The nnU-Net and its atlas-augmented variant (nnU-Net(A)) both achieved the highest mean DSC of 0.71 for Subtask 2a, outperforming all other methods. Atlas augmentation further improved nnU-Net's boundary accuracy, yielding the lowest HD95 and ASSD scores of 2.08 and 0.70, respectively. The smallest RVE (0.13) was observed in both VNet and atlas-augmented VNet (VNet(A)), indicating accurate volumetric estimates despite lower overlap and boundary metrics.

Table 1. Subtask 2a—Average results on the validation set with hippocampus labels (DSC, HD, HD95 ASSD, RVE). Best in **bold**, second best <u>underlined</u>.

Method	DSC $\uparrow$	HD $\downarrow$	HD95 $\downarrow$	ASSD $\downarrow$	RVE $\downarrow$
VNet	<u>0.63±0.16</u>	6.46±7.71	2.55±1.37	0.94±0.80	**0.13±0.11**
VNet(A)	0.63±0.18	6.57±7.71$^{(\uparrow 0.11)}$	2.59±1.57$^{(\uparrow 0.04)}$	1.00±1.01$^{(\uparrow 0.06)}$	<u>0.13±0.12</u>
MS2+2D	0.58±0.15	8.86±8.22	3.67±1.49	1.09±0.56	0.22±0.14
MS2+3D	0.60±0.17	7.39±8.06	2.97±1.67	1.04±0.87	0.14±0.08
MS2+3D(A)	0.59±0.16$^{(\downarrow 0.01)}$	7.40±7.93$^{(\uparrow 0.01)}$	2.96±1.70	0.97±0.58	0.16±0.11$^{(\uparrow 0.02)}$
nnU-Net	**0.71±0.18**	**5.90±7.96**	<u>2.13±1.64</u>	<u>0.76±0.88</u>	0.16±0.10
nnU-Net(A)	**0.71±0.18**	<u>5.97±7.99$^{(\uparrow 0.07)}$</u>	**2.08±1.41**	**0.70±0.76**	0.15±0.07

5.2   Results 2b: Basal Ganglia

For Subtask 2b, nnU-Net and nnU-Net(A) again led the cohort with a mean DSC of 0.86. Atlas augmentation enhanced all distance and volume metrics, achieving an HD of 0.45, HD95 of 1.57, ASSD of 0.45, and RVE of 0.08. The VNet and VNet(A) models ranked second in DSC (0.83) but exhibited higher HD95 (1.86) and ASSD (0.59), underscoring the benefits of atlas priors when integrated with the nnU-Net framework.

Table 2. Subtask 2b—Average results on the validation set with basal ganglia labels (DSC, HD, HD95, ASSD, RVE). Best in **bold**, second best <u>underlined</u>.

Method	DSC ↑	HD ↓	HD95 ↓	ASSD ↓	RVE ↓
VNet	<u>0.83±0.04</u>	3.43±0.66	1.85±0.61	0.58±0.16	0.11±0.06
VNet(A)	<u>0.83±0.04</u>	3.35±0.62 $^{(\downarrow 0.08)}$	1.86±0.56 $^{(\uparrow 0.01)}$	0.59±0.15 $^{(\uparrow 0.01)}$	<u>0.10±0.05</u> $^{(\downarrow 0.01)}$
MS2+2D	0.79±0.03	6.04±2.51	2.51±0.60	0.76±0.13	0.12±0.04
MS2+3D	0.81±0.04	4.28±0.69	2.17±0.52	0.68±0.14	0.14±0.09
MS2+3D(A)	0.81±0.03	4.01±0.96 $^{(\downarrow 0.27)}$	2.08±0.56 $^{(\downarrow 0.09)}$	0.66±0.16 $^{(\downarrow 0.02)}$	0.10±0.07 $^{(\downarrow 0.04)}$
nnU-Net	**0.86±0.05**	<u>2.97±0.89</u>	<u>1.65±0.83</u>	<u>0.45±0.22</u>	**0.08±0.04**
nnU-Net(A)	**0.86±0.05**	**2.95±0.69** $^{(\downarrow 0.02)}$	**1.57±0.54** $^{(\downarrow 0.08)}$	**0.45±0.19**	**0.08±0.04**

5.3   Atlas Augmentation Ablation Study

To quantify the impact of atlas priors, we report the change in each metric for VNet, MS2+3D, and nnU-Net in both subtasks. These changes are presented in the Tables 1 and 2 with red and green text, indicating either the improvement or deterioration with respect to the scores.

Overall, in Subtask 2a, atlas augmentation slightly degrades the VNet, has negligible or slightly negative effect on MS2+3D, and yields modest but consistent boundary and volume gains for nnU-Net.

In Subtask 2b, all augmented models exhibit net improvements in distance and volume metrics, with nnU-Net(A) achieving the most balanced gains and MS2+3D(A) showing the largest absolute reductions in HD and RVE.

These results confirm that atlas priors confer the greatest benefit when integrated with nnU-Net, particularly for hippocampal segmentation (Subtask 2a), while providing more uniform improvements across models in basal ganglia delineation (Subtask 2b).

6   Discussion

Our results show that nnU-Net(A)—which combines nnU-Net's self-configuring pipeline with simple atlas channel augmentation—yields the best overall performance on uLF pediatric MRI, improving boundary and volume metrics despite low contrast and high noise. VNet and MedSAM2 variants benefit less from priors, suggesting that flexible preprocessing and architecture search (as in nnU-Net) better exploit external anatomical information.

Interestingly, atlas augmentation led to relatively greater gains for basal ganglia segmentation than for the hippocampus. This likely reflects the fact that probabilistic atlases encode more robust and spatially consistent information for larger, more homogeneous structures like the basal nuclei, whereas the smaller and more variable hippocampi may receive less precise guidance, highlighting an inherent class-size and prior-information imbalance. However, this is worth mentioning that the registration and atlas extraction can for some images double or even triple the time of final segmentation. Since the time constraint for the predictions was not a part of limitations in the challenge, the atlas augmented models were preferred in the final submissions.

Limitations include the use of adult/older-child atlases that may misalign with our young cohort, and we have not explicitly modeled the full range of low-field artifacts—such as spatially varying SNR, field inhomogeneities, and motion artifacts—that can degrade segmentation. While our models are trained on actual 0.064 T scans and show stable performance on the validation set, their robustness to further SNR degradation or synthetic noise remains untested. Future work should incorporate noise- and bias-field augmentation to simulate variable low-field conditions, evaluate model stability under reduced SNR, and explore age-matched templates, uncertainty estimation, and targeted fine-tuning (e.g., via LoRA [29]) to adapt pretrained networks more fully to challenging low-field data. Overall, our study highlights the synergistic value of automated model configuration and atlas priors for robust segmentation in resource-limited imaging settings.[1]

Acknowledgments. This work is supported by Fraunhofer MEVIS DeLoRI research on the low field MRI (KL) and the German Research Foundation DFG (ZC (STE2802/5–1)).

Disclosure of Interests. The authors have no competing interests to declare that are relevant for this article.

References

1. Hori, M., Hagiwara, A., Goto, M., Wada, A., Aoki, S.: Low-field magnetic resonance imaging. Invest. Radiol. **56**(11), 669–679 (2021)
2. Kersting-Sommerhoff, B., Hof, N., Lenz, M., Gerhardt, P.: MRI of peripheral joints with a low-field dedicated system: a reliable and cost-effective alternative to high-field units? Eur. Radiol. **6**(4), 561–565 (1996)
3. Sarracanie, M., Salameh, N.: Low-Field MRI: how low can we go? a fresh view on an old debate. Front. Phys., 8 (2020)
4. Arnold, T.C., Freeman, C.W., Litt, B., Stein, J.M.: Low-field MRI: clinical promise and challenges. J. Magnetic Res. Imag. **57**(1), 25–44 (2023)

---

[1] Team ID: https://www.synapse.org/Team:3551168.

5. Ayde, R., Vornehm, M., Zhao, Y., Knoll, F., Wu, E.X., Sarracanie, M.: MRI at low field: a review of software solutions for improving SNR. NMR Biomed. **38**(1), e5268 (2025)

6. Salat, D.H., Lee, S.Y., Van der Kouwe, A.J., Greve, D.N., Fischl, B., Rosas, H.D.: Age-associated alterations in cortical gray and white matter signal intensity and gray to white matter contrast. Neuroimage **48**(1), 21–28 (2009)

7. Zhou, W., Li, J., Wang, X., Wang, Y., Lyu, M.: Infant hippocampal segmentation in ultra-low-field mri using external datasets with diverse field strengths. In: Lepore, N., Linguraru, M.G., eds., Low Field Pediatric Brain Magnetic Resonance Image Segmentation and Quality Assurance, pp. 28–37, Cham. Springer Nature Switzerland (2025)

8. Tapp, A., et al.: Quality assurance and hippocampal segmentation on low-field pediatric magnetic resonance images. In: Lepore, N., Linguraru, M.G. (eds.) Low Field Pediatric Brain Magnetic Resonance Image Segmentation and Quality Assurance, pp. 63–75. Springer Nature Switzerland, Cham (2025)

9. Sundaresan, V., Dinsdale, N.K.: Automated quality assessment using appearance-based simulations and hippocampus segmentation on low-field paediatric brain MR images. In: Lepore, N., Linguraru, M.G., (eds.), Low Field Pediatric Brain Magnetic Resonance Image Segmentation and Quality Assurance, pp. 41–52. Springer Nature Switzerland, Cham (2025)

10. Peiris, H., Chen, Z.: Bilateral hippocampi segmentation in low field MRIs using mutual feature learning via dual-views. In: Lepore, N., Linguraru, M.G. (eds.) Low Field Pediatric Brain Magnetic Resonance Image Segmentation and Quality Assurance, pp. 15–27. Springer Nature Switzerland, Cham (2025)

11. Milletari, F., Navab, N., Ahmadi, S.A.: V-Net: fully convolutional neural networks for volumetric medical image segmentation. arXiv:1606.04797 (2016)

12. Ma, J., et al.: MedSAM2: segment anything in 3d medical images and videos. arXiv:2504.03600 (2025)

13. Isensee, F., et al.: NNU-Net: self-adapting framework for U-Net-based medical image segmentation. arXiv:1809.10486 (2018)

14. Mazziotta, J.C., Toga, A.W., Evans, A., Fox, P., Lancaster, J.: A probabilistic atlas of the human brain: theory and rationale for its development. The international consortium for brain mapping (ICBM). NeuroImage, **2**(2), 89–101 (1995)

15. Avants, B., Tustison, N.J., Song, G.: Advanced normalization tools: V1.0. Insight J. (2009)

16. Hyperfine, Inc. and the Swoop® Portable MR Imaging® System

17. Low Field Pediatric Brain Magnetic Resonance Image Segmentation and Quality Assurance: First MICCAI Challenge, LISA 2024, Held in Conjunction with MICCAI 2024, Marrakesh, Morocco, October 10, 2024, Proceedings | SpringerLink

18. Kim, H., Seo, J., Ryu, S., Park, J.H., On, S., Choi, J.: Axis-guided quality assessment and multi-label hippocampal and ventricular segmentation in low-resolution pediatric brain MRI. In: Lepore, N., Linguraru, M.G. (eds.) Low Field Pediatric Brain Magnetic Resonance Image Segmentation and Quality Assurance, pp. 53–62. Springer Nature Switzerland, Cham (2025)

19. Jorge Cardoso, M., et al.: MONAI: an open-source framework for deep learning in healthcare. arXiv:2211.02701 (2022)

20. Pérez-García, F., Sparks, R., Ourselin, S.: TorchIO: a python library for efficient loading, preprocessing, augmentation and patch-based sampling of medical images in deep learning. Comput. Methods Programs Biomed., **208**, 106236 (2021). arXiv:2003.04696

21. Nikolaos, A.: Deep learning in medical image analysis: a comparative analysis of multi-modal brain-MRI segmentation with 3D deep neural networks. Master's thesis, University of Patras (2019)
22. Ravi, N., et al.: SAM 2: segment anything in images and videos. arXiv:2408.00714 (2024)
23. Bui, N., et al.: SAM3D: segment anything model in volumetric medical images. arXiv:2309.03493 (2023)
24. Pauli, W.M., Nili, A.N., Tyszka, J.M.: A high-resolution probabilistic in vivo atlas of human subcortical brain nuclei. Sci. Data 5(1), 180063 (2018)
25. Makris, N., et al.: Decreased volume of left and total anterior insular lobule in schizophrenia. Schizophrenia Res. 83(2), 155–171 (2006)
26. Liu, H., Nie, D., Yang, J., Wang, J., Tang, Z.: A new multi-atlas based deep learning segmentation framework with differentiable atlas feature warping. IEEE J. Biomed. Heal. Inf. 28(3), 1484–1493 (2024)
27. Yeghiazaryan, V., Voiculescu, I.: Family of boundary overlap metrics for the evaluation of medical image segmentation. J. Med. Imaging 5(1), 015006 (2018)
28. Maier-Hein, L., et al.: Bias: transparent reporting of biomedical image analysis challenges. Med. Image Anal. 66, 101796 (2020)
29. Hu, E.J., et al.: LoRA: low-rank adaptation of large language models (2021). arXiv:2106.09685

Automated Pediatric Brain Hippocampal and Basal Ganglia Segmentation in Ultra-low Field Magnetic Resonance Images

Toufiq Musah[1(✉)], Philip Nkwam[2], and Ajay Sharma[3]

[1] Department of Computer Engineering, Kwame Nkrumah University of Science and Technology, Kumasi, Ghana
`tmusah@st.knust.edu.gh`
[2] Department of Radiography, Faculty of Health Professions, College of Medicine, University of Lagos, Lagos, Nigeria
[3] Laboratory for Accessible MRI, Division of Cancer Imaging Research, Radiology and Radiological Sciences, Johns Hopkins University School of Medicine, Baltimore, USA

Abstract. The automated segmentation of brain structures, such as the hippocampus and basal ganglia, from ultra-low field (0.064T) neonatal magnetic resonance imaging (MRI) is important for clinical diagnosis and neurodevelopmental research. Manual segmentation is time-consuming and subject to inter- and intra-observer variability. The unique challenges of ultra-low field MRI, including low signal to noise ratio and low spatial resolution make automated segmentation a difficult task. In this study, the challenge of automated segmentation of the basal ganglia and hippocampus in pediatric ultra-low field MRI is addressed. The approach builds on MedNeXt, a transformer-inspired, fully convolutional encoder-decoder architecture, trained with the nnU-Net pipeline. To address the class imbalance inherent in segmenting small structures, a combined Focal-Dice-CrossEntropy loss function was employed. The method was evaluated using the Dice Similarity Coefficient (DSC), 95th Percentile Hausdorff Distance (HD95), Average Symmetric Surface Distance (ASSD), and Relative Volume Error (RVE). The results show an average DSC of 0.69 ± 0.18 for hippocampal segmentation and an average DSC of 0.85 ± 0.06 for basal ganglia segmentation. The method demonstrated better performances in segmenting the basal ganglia in the ultra-low field images as compared to the hippocampus.

Keywords: Ultra-low field MRI · Neonatal brain · Hippocampus · Basal ganglia · Automated segmentation · MedNeXt · LISA 2025

1 Introduction

The automated segmentation of the hippocampus and basal ganglia in ultra-low field neonatal magnetic resonance imaging (MRI) is a challenging task. This is

© The Author(s) 2026
N. Lepore and M. G. Linguraru (Eds.): LISA 2025, LNCS 16411, pp. 98–105, 2026.
https://doi.org/10.1007/978-3-032-14417-1_9

due to low signal to noise ratio and spatial resolution of the images, compared to High-field MRIs, with their superior image quality that are typically used for such task [4]. The hippocampus and basal ganglia play vital roles in neurodevelopment and are associated with various neurological disorders, making automated segmentation of these structures in ultra low field MRI important for both clinical diagnosis and research [6,8]. Manual segmentation methods are often time consuming and inconsistent, suffering from inter- and intra-observer variability [5], prompting the need for automated methods [3,23]. Hippocampal and basal ganglia segmentation is also vital for early detection of cerebral palsy and autism, which can have long-lasting implications for cognitive function and quality of life [15,18]. As such, the integration of automated techniques in ultra-low field MRI facilitates better clinical outcomes, while enhancing the understanding of the underlying mechanisms of brain health [6].

Recent studies have explored the segmentation of the hippocampus in ultra-low field MRI, including [20] where they first enhance the ultra-low field MRI to a high-field variant in a process referred to as Super-Field [21], before performing segmentation using nnU-Net [7]. Other studies have applied the nnU-Net framework directly to the segmentation of ultra-low field MRI [10,22]. A linear registration approach and a 3D-UNet with priors to correct under-segmentation was introduced by [19]. The LoFiHippSeg architecture proposed by [16] introduces a dual-view deep learning framework, using VNet as the segmentation backbone, designed to learn complementary features from low-field MRI for hippocampal segmentation. While architectures such as nnU-Net [7] and MedNeXt [17] were originally trained on adult populations data, their direct application to neonatal imaging is not straightforward. Neonatal brains differ from adult brains in tissue contrast, size, and developmental variability, which may lead to suboptimal performance when said models are used without adaptation.

In this study, we address the challenge of segmenting the hippocampus and basal ganglia from ultra-low field MRI, building on MedNeXt [17], a transformer-inspired fully convolutional encoder-decoder architecture based on the ConvNeXt [13] architecture. To address the significant challenge of class imbalance of foreground vs. background inherent in segmenting small anatomical structures like the hippocampus and basal ganglia where target regions occupy a little percentage of the brain volume, we employ a composite loss function combining Focal loss, Dice loss, and Cross-Entropy loss. This approach is designed to enhance segmentation precision by giving greater weight to misclassified foreground pixels, thereby improving the model's ability to accurately delineate the target anatomies [9,12]. To the best of our knowledge, this study represents the first application of MedNeXt to neonatal brain structure segmentation in ultra-low field MRI, using a composite loss that prioritizes small-structure accuracy.

2   Methods

2.1   Data

This study used the LISA 2025 Task 2 dataset, comprising 79 training samples and 12 validation samples of low-field (0.064T Hyperfine) MRI scans. High-field T2-weighted scans served as the anatomical reference for expert-annotated ground-truth segmentations of left and right bilateral hippocampi, ventricles, and basal ganglia. All data handling and preprocessing were performed using the default nnU-Net pipeline [7] (Fig. 1).

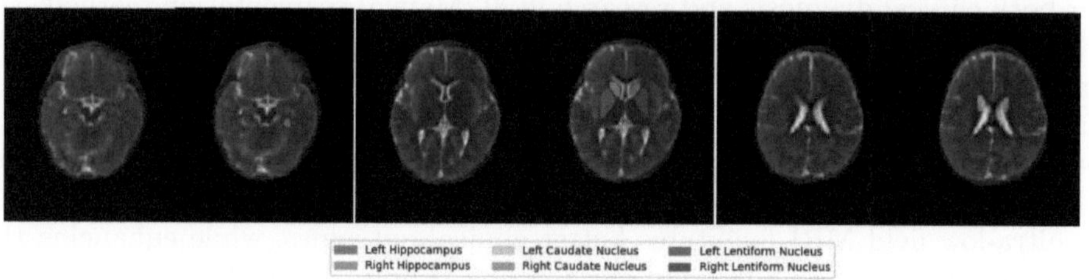

Fig. 1. Annotated ultra-low field MRI slices showing bilateral hippocampi and basal ganglia segmentations.

2.2   Segmentation of Hippocampi and Basal Ganglia

The proposed method for segmenting the hippocampi and basal ganglia employs MedNeXt [17], a transformer-inspired large-kernel segmentation network. MedNeXt introduces several key innovations: (1) a fully ConvNeXt-based [13] 3D encoder – decoder architecture designed for volumetric medical image segmentation, (2) residual ConvNeXt upsampling and downsampling blocks that preserve semantic richness across scales, (3) a novel technique for progressively increasing kernel sizes by upsampling small-kernel networks, preventing performance saturation on limited medical datasets, and (4) compound scaling across depth, width, and kernel size.[1]

This model was trained within the widely adopted nnU-Net pipeline [7] for 500 epochs on an NVIDIA A100 GPU. To address the challenge of segmenting relatively small anatomical structures of the left and right hippocampi and basal ganglia within a large background, we augmented the standard Dice – Cross Entropy loss with a Focal loss component, as shown in Eq. 1. Focal loss down-weights well-classified background pixels and assigns greater importance to hard-to-classify foreground pixels, improving sensitivity to small structures that are easily overlooked by conventional loss functions [11]. This composite formulation helps mitigate the high background-to-foreground pixel ratio and reduces bias toward the background.

$$\mathcal{L}_{\text{total}} = 0.25\,\mathcal{L}_{\text{Dice-CE}} + 0.75\,\mathcal{L}_{\text{Focal}} \tag{1}$$

MedNeXt was chosen for its demonstrated superiority in multiple biomedical image segmentation tasks, particularly in neuroimage segmentation [1,2,14]. Our implemented variant remains computationally efficient, containing 30.8 million parameters.

---

[1] MedNeXt is publicly available at: https://github.com/MIC-DKFZ/MedNeXt.

2.3   Evaluation

To assess the performance of our method, we evaluated it on the 12 validation samples from the LISA 2025 Task 2 dataset. The evaluation metrics included the Dice Similarity Coefficient (DSC), Hausdorff Distance (HD), 95th Percentile Hausdorff Distance (HD95), Average Symmetric Surface Distance (ASSD), and Relative Volume Error (RVE). These metrics were computed for both the left and right hippocampi (Task 2a), as well as the left and right basal ganglia (Task 2b), which further includes the caudate nucleus and lentiform nucleus. The mean and standard deviation of each metric were calculated across the validation set to provide a good measure of the method's performance (Fig. 2).

3   Results and Discussion

The performance of the proposed model for segmenting the hippocampi and basal ganglia on the LISA 2025 validation set is shown in Tables 1 and 2.

Table 1. Segmentation performance for hippocampi (Task 2a) on the 12-case LISA 2025 validation set. Values are reported as mean ± standard deviation. We compare these to the average of the uLF nnU-Net as a baseline.

Structure	DSC(↑)	HD(↓)	HD95(↓)	ASSD(↓)	RVE(↓)
L. Hippocampus	0.64 ± 0.23	8.71 ± 16.34	2.71 ± 2.22	1.04 ± 1.38	0.20 ± 0.13
R. Hippocampus	0.74 ± 0.14	3.49 ± 1.23	1.75 ± 0.76	0.53 ± 0.35	0.11 ± 0.08
Average	**0.69** ± 0.18	**6.10** ± 8.14	**2.23** ± 1.40	**0.79** ± 0.86	**0.15** ± 0.10
uLF nnU-Net [20]	0.61 ± 0.27	13.55 ± 15.77	8.43 ± 13.08	4.03 ± 10.45	0.16 ± 0.11

For Task 2a, the segmentation of the hippocampi, the model achieved an average Dice Similarity Coefficient (DSC) of 0.69 ± 0.18. A notable difference was observed between the two structures, with the right hippocampus (DSC: 0.74 ± 0.14) showing a higher and more consistent performance than the left hippocampus (DSC: 0.64 ± 0.23). The error metrics, including the 95th Percentile Hausdorff Distance (HD95), Average Symmetric Surface Distance (ASSD), and Relative Volume Error (RVE), followed a similar trend, with lower values for the right hippocampus, indicating better surface and volume overlap.

Segmentation of the basal ganglia in Task 2b demonstrated a significantly higher performance. The average DSC across all four structures (left and right caudate, left and right lentiform nuclei) was 0.85 ± 0.06. The results for the individual structures were consistently high, with the right lentiform nucleus achieving the highest DSC of 0.87 ± 0.06. The standard deviations for all metrics were considerably lower than those for the hippocampi, indicating greater model stability and more robust performance across the validation set. The error metrics (HD, HD95, ASSD, RVE) for the basal ganglia were also lower than those for the

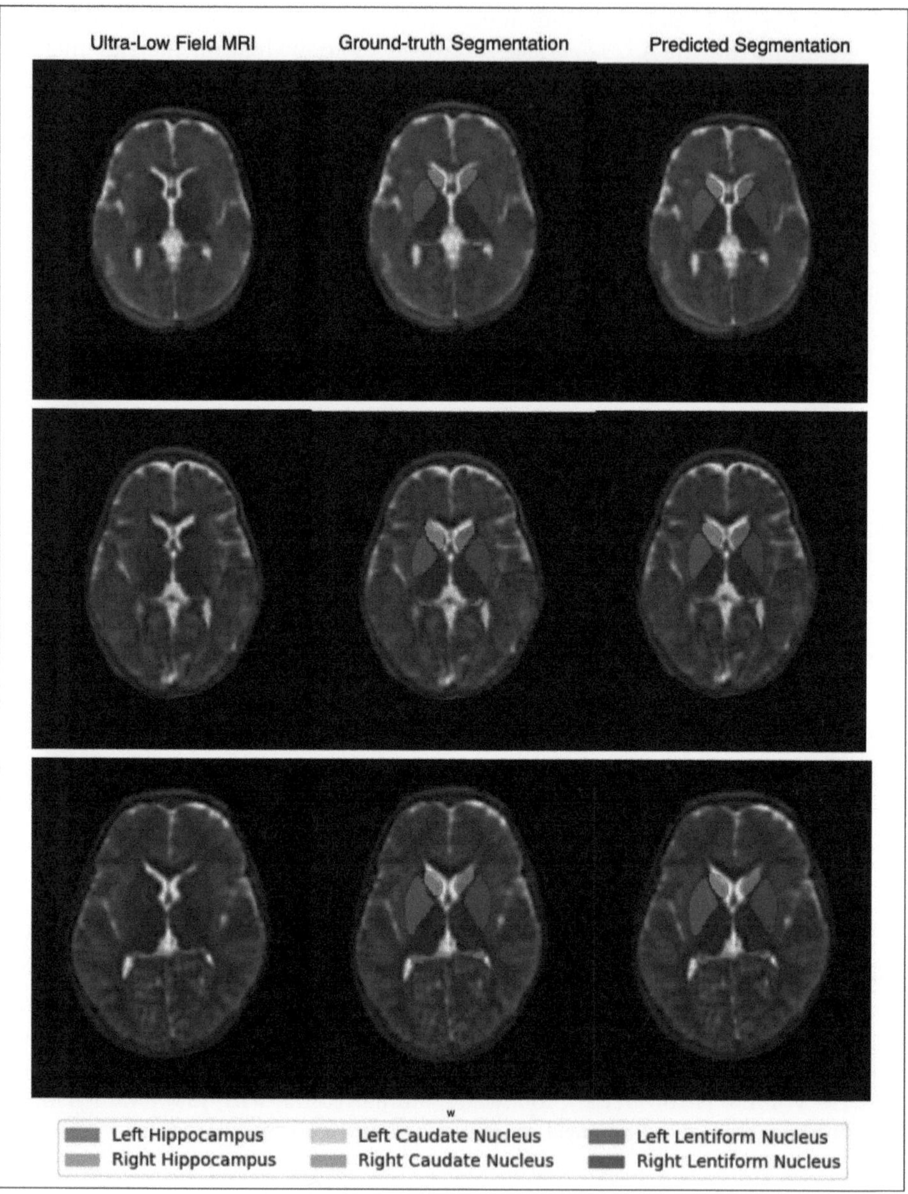

Fig. 2. Qualitative results on ultra-low field (ULF) MRI. Each row shows a representative axial slice from a different subject. Columns show (left) the raw ULF MRI, (middle) the ground-truth manual segmentations, and (right) the corresponding model predictions.

Table 2. Segmentation performance for basal ganglia (Task 2b) on the 12-case LISA 2025 validation set. Values are reported as mean ± standard deviation.

Structure	DSC(↑)	HD(↓)	HD95(↓)	ASSD(↓)	RVE(↓)
Left Caudate	0.83 ± 0.07	3.84 ± 1.22	1.75 ± 0.76	0.47 ± 0.24	0.09 ± 0.08
Right Caudate	0.84 ± 0.08	2.73 ± 0.99	1.50 ± 0.79	0.43 ± 0.28	0.06 ± 0.06
Left Lentiform	0.85 ± 0.06	3.21 ± 1.06	2.15 ± 1.26	0.62 ± 0.28	0.10 ± 0.09
Right Lentiform	0.87 ± 0.06	2.96 ± 1.08	1.82 ± 1.18	0.51 ± 0.27	0.09 ± 0.06
Average	**0.85 ± 0.06**	**3.18 ± 0.86**	**1.81 ± 0.90**	**0.51 ± 0.24**	**0.08 ± 0.04**

hippocampi, suggesting that the model predicted more accurate and consistent segmentations for these structures.

Testing Phase Performance. The official LISA 2025 testing phase results are summarized in Table 3. The trends were consistent with the validation phase. For Task A (hippocampi segmentation), the model achieved a mean Dice score of 0.54 ± 0.21, reflecting the inherent difficulty of delineating small, low-contrast structures. For Task B (basal ganglia segmentation), the model maintained strong performance with a mean Dice score of 0.80 ± 0.10 and lower surface distance errors.

Table 3. Official LISA 2025 testing phase results for Tasks 2a and 2b

Task	DSC(↑)	HD(↓)	HD95(↓)	ASSD(↓)	RVE(↓)
Task 2a	0.54 ± 0.21	4.76 ± 2.10	3.22 ± 1.63	1.13 ± 0.62	0.20 ± 0.11
Task 2b	0.80 ± 0.10	3.89 ± 0.99	2.28 ± 0.89	0.72 ± 0.40	0.19 ± 0.15

4   Conclusion

In this work, we addressed the challenges of segmenting pediatric brain structures from ultra-low field MRI scans as part of the LISA 2025 challenge. We proposed a segmentation framework based on the MedNeXt model, trained within the nnU-Net pipeline and using a modified Dice-CE loss with a Focal loss component to handle class imbalance. The model achieved an average Dice of 0.69 and 0.85 for the hippocampi and basal ganglia respectively. The higher performance on the basal ganglia suggests that these structures may have more distinct image characteristics or are less susceptible to the challenges presented by ultra-low field MRI, such as lower contrast, compared to the hippocampi.

While our approach shows promising results, it is important to acknowledge its limitations. The absence of a comprehensive comparison against other well-established methods, such as variants of MedNeXt without the addition of the

Focal loss is a major concern. This lack of benchmarking makes it difficult to definitively assess the specific contribution of our proposed modifications.

Future work will focus on addressing the identified limitations, by conducting a thorough ablation study to evaluate the impact of the Focal loss and by benchmarking our method against state-of-the-art architectures to better understand its relative strengths and weaknesses. We also plan to investigate the reasons for the performance disparity between the hippocampi and basal ganglia, exploring whether this is due to inherent anatomical differences, image characteristics, or a need for more specialized model architectures.

References

1. Adhikari, B., et al.: Parameter-efficient fine-tuning for improved convolutional baseline for brain tumor segmentation in sub-Saharan Africa adult glioma dataset. arXiv preprint arXiv:2412.14100 (2024)
2. Ankomah, C.T., et al.: How we won brats-SSA 2025: Brain tumor segmentation in the sub-saharan african population using segmentation-aware data augmentation and model Ensembling. arXiv preprint arXiv:2510.03568 (2025)
3. Billot, B., Greve, D.N., Van Der Kouwe, A., Fischl, B., Iglesias, J.E.: A learning strategy for contrast-agnostic MRI segmentation. Med. Image Anal. **72**, 102102 (2021)
4. Ciceri, T., Squarcina, L., Giubergia, A., Bertoldo, A., Brambilla, P., Peruzzo, D.: Review on deep learning fetal brain segmentation from magnetic resonance images. Artif. Intell. Med. **143**, 102608 (2023)
5. Collier, D.C., et al.: Assessment of consistency in contouring of normal-tissue anatomic structures. J. Appl. Clin. Med. Phys. **4**(1), 17–24 (2003)
6. Gilmore, J.H., Knickmeyer, R.C., Gao, W.: Imaging structural and functional brain development in early childhood. Nat. Rev. Neurosci. **19**, 123–137 (2018)
7. Isensee, F., et al.: nnu-net: Self-adapting framework for u-net-based medical image segmentation. arXiv preprint arXiv:1809.10486 (2018)
8. Jacob, F.D., et al.: Fetal hippocampal development: analysis by magnetic resonance imaging volumetry. Pediatr. Res. **69**(5), 425–429 (2011)
9. Jadon, S.: A survey of loss functions for semantic segmentation. Image Video Process. **2020**, 1–11 (2020)
10. Kim, H., et al.: Axis-guided quality assessment and multi-label hippocampal and ventricular segmentation in low-resolution pediatric brain MRI. In: MICCAI Challenge on Low Field Pediatric Brain Magnetic Resonance Image Segmentation and Quality Assurance, pp. 53–62. Springer Nature Switzerland Cham (2024)
11. Lin, T.Y., Goyal, P., Girshick, R., He, K., Dollár, P.: Focal loss for dense object detection. In: Proceedings of the IEEE International Conference on Computer Vision, pp. 2980–2988 (2017)
12. Lin, T.Y., Goyal, P., Girshick, R., He, K., Dollár, P.: Focal loss for dense object detection. IEEE Trans. Pattern Anal. Mach. Intell. **42**(2), 318–327 (2020)
13. Liu, Z., et al.: A convnet for the 2020s. In: Proceedings of the IEEE/CVF Conference on Computer Vision and Pattern Recognition, pp. 11976–11986 (2022)
14. Parida, A., et al.: Adult glioma segmentation in sub-saharan africa using transfer learning on stratified finetuning data. arXiv preprint arXiv:2412.04111 (2024)

15. Payne, A., et al.: Predicting neurodevelopmental outcomes in infants using neonatal MRI: A review of the current state and future directions. NeuroImage: Clinical **30**, 102639 (2021)
16. Peiris, H., Chen, Z.: Bilateral hippocampi segmentation in low field MRIs using mutual feature learning via dual-views. In: MICCAI Challenge on Low Field Pediatric Brain Magnetic Resonance Image Segmentation and Quality Assurance, pp. 15–27. Springer Nature Switzerland Cham (2024)
17. Roy, S., et al.: Mednext: transformer-driven scaling of convnets for medical image segmentation. In: International Conference on Medical Image Computing and Computer-Assisted Intervention, pp. 405–415. Springer (2023)
18. Shi, F., Xia, S., Lin, W., Gilmore, J.H., Shen, D.: Discriminative analysis of early infant brain functional connectivity. Neuroimage **82**, 180–191 (2013)
19. Sundaresan, V., Dinsdale, N.K.: Automated quality assessment using appearance-based simulations and hippocampus segmentation on low-field Paediatric brain MR images. In: MICCAI Challenge on Low Field Pediatric Brain Magnetic Resonance Image Segmentation and Quality Assurance, pp. 41–52. Springer (2024)
20. Tapp, A., et al.: Quality assurance and hippocampal segmentation on low-field pediatric magnetic resonance images. In: MICCAI Challenge on Low Field Pediatric Brain Magnetic Resonance Image Segmentation and Quality Assurance, pp. 63–75. Springer Nature Switzerland Cham (2024)
21. Tapp, A., et al.: Super-field MRI synthesis for infant brains enhanced by dual channel latent diffusion. In: International Conference on Medical Image Computing and Computer-Assisted Intervention, pp. 444–454. Springer (2024)
22. Zhou, W., Li, J., Wang, X., Wang, Y., Lyu, M.: Infant hippocampal segmentation in ultra-low-field mri using external datasets with diverse field strengths. In: MICCAI Challenge on Low Field Pediatric Brain Magnetic Resonance Image Segmentation and Quality Assurance, pp. 28–37. Springer Nature Switzerland Cham (2024)
23. Zhou, Y., Li, X., Liang, X., Wang, Y., Yang, X.: Automatic hippocampus segmentation using deep learning in magnetic resonance imaging: A review. Comput. Biol. Med. **133**, 104375 (2021)

Tasks 1, 2a and 2b Combined

Application of Vision Transformers to Multi-task Learning in the LISA 2025 MRI Challenge

Tian Song[1] and Dou Jiaqi[2]

[1] Philips Healthcare, Beijing, China
[2] Hepatopancreatobiliary Center, Beijing Tsinghua Changgung Hospital, School of Clinical Medicine, Tsinghua Medicine, Tsinghua University, Beijing, China
doujq@mail.tsinghua.edu.cn

Abstract. Transformer-based architectures are increasingly used in medical image analysis to support diverse tasks under a unified framework. Our solution for the LISA 2025 challenge addresses both image quality classification (Task I) and semantic segmentation (Task II). For Task I, we introduced a slice-wise strategy based on a Vision Transformer (ViT) pre-trained on ImageNet. The model processes 3D MRI volumes decomposed into 2D slices, each carrying the original volume's quality label. Predictions are combined by selecting the maximum value across slices for each quality category. The ViT encoder remained frozen throughout training, with updates limited to the classification layer. In Task II, a UNETR architecture was applied, incorporating encoder weights pre-trained on SAM-Med 3D. Training involved two stages: initial optimization of the decoder with a fixed encoder, followed by full model fine-tuning using Low-Rank Adaptation (LoRA). In the testing stage, Our approach achieved a weighted F1 score of 0.781 for quality assessment, and average Dice scores of 0.58 and 0.81 for hippocampal and basal ganglia segmentation, respectively. These outcomes highlight the flexibility and effectiveness of transformer-based models in multi-task medical image analysis. Our code for Task 1 has been made openly available at https://github.com/RimeT/lisa2025_task1_teamCGP.

Keywords: Transformers · Deep Learning · Image Classification · Semantic Segmentation

1 Introduction

Over the past decade, deep learning for medical image analysis has been predominantly dominated by Convolutional Neural Networks (ConvNets) [1]. To achieve optimal performance on specific tasks, different network architectures were typically employed for classification and segmentation—for instance, ResNet [1] for classification and U-Net [2] for medical image segmentation. This approach often led to practical inconveniences in code implementation when attempting to transfer learning or share encoder knowledge across tasks. In particular, the encoder structures of classification networks and segmentation models such as U-Net are usually designed differently, and the naming conventions of convolution blocks vary significantly between architectures. As a result, reusing

© The Author(s) 2026
N. Lepore and M. G. Linguraru (Eds.): LISA 2025, LNCS 16411, pp. 109–118, 2026.
https://doi.org/10.1007/978-3-032-14417-1_10

pretrained weights between models becomes cumbersome. Furthermore, the inherent differences between 2D and 3D convolutional operations make parameter sharing even more challenging, complicating the development of unified frameworks.

The advent of the Vision Transformer (ViT) [3] has begun to break down these barriers between data structures and tasks. With the emergence of methods like CLIP [4], it has been demonstrated that a single network can handle different data modalities effectively. The Transformer encoder processes input and output as sequences of features, which facilitates the sharing of backbone networks between 2D and 3D data. Specifically, while 2D data embedding is typically achieved using a 2D convolution, 3D embedding may involve 3D convolutions or 2D convolutions applied slice-wise; notably, studies such as ViViT [5] have shown that 3D convolutions can perform comparably to 2D approaches in certain contexts.

Several Transformer-based architectures are now being applied to various medical imaging tasks: ViT3D for classification and UNETR [6] for semantic segmentation. Large-scale pre-training has proven crucial for fine-tuning Transformer networks effectively [3, 4, 7]. The availability of large-scale pre-trained models for MRI is gradually expanding, with initiatives like SAM-Med 3D [8]—which uses a ViT with an 3D embedding size of 384 for instance segmentation. Other notable examples include models trained on increasingly extensive and diverse medical image datasets.

The Low-field pediatric brain magnetic Resonance Image Segmentation and Quality Assurance (LISA) challenge was established to catalyze the development of automated tools for portable, low-field MRI in resource-limited settings. The inaugural LISA 2024 challenge [9] set a significant precedent, garnering participation from 36 teams globally and yielding a diverse set of advanced methodologies for its two core tasks. For Task I (Image Quality Assessment), the winning solution was based on a DenseNet architecture enhanced with appearance-based transformations to simulate artifacts and address class imbalance. This success was complemented by other innovative approaches, such as the Multi-Label MambaOut network which used gated convolutional blocks. For Task II (Hippocampal Segmentation), the winning team used the nnUNet framework, while other leading methods included a dual-view architecture (LoFiHippSeg) for mutual feature learning. Building upon this solid foundation, the LISA 2025 challenge continues to refine these objectives for quality assurance and hippocampal segmentation, while further advancing the field by introducing the segmentation of the basal ganglia.

In the LISA 2025 challenge, which comprises two tasks, we used ViT as the backbone for both. Task I involved image quality assessment, with 7 categories each having 3 quality levels, and was addressed using a 2D ViT-base model pre-trained on ImageNet. For Task II, which focused on semantic segmentation, we used a UNETR architecture, initializing its encoder with pre-trained weights from SAM-Med 3D's ViT encoder.

2 Methods

2.1 Task I Quality Assessment

For Task I (image quality assessment), we simulated the human evaluation process by training a quality assessment model in a slice-wise manner. The original 3D volume was divided into 2D axial slices. During training, each slice was assigned the label of its corresponding parent 3D volume. Converting 3D volumes into 2D slices not only significantly expands the effective training dataset size but also facilitates model fine-tuning and helps mitigate overfitting.

2.2 Image Preprocessing for Task I

During the preprocessing of 3D MRI images, we preserved the original inter-slice and intra-slice resolutions without performing any respacing. Z-normalization was first applied to standardize the voxel values of the 3D volumes. Subsequently, the normalized volumes were split into individual 2D slices. Before input into the model, all 2D slices were resized to a uniform dimension of 224×224 pixels using the "longest" resize mode (maintaining aspect ratio by padding the shorter side) to ensure consistent batch dimensions. Data augmentation was applied exclusively online during the training phase. We used only spatial augmentations, specifically random flipping and random affine transformations. Notably, no intensity-based augmentations (such as adjustments to contrast, or gamma) were used. This approach was chosen to preserve the original voxel intensity distributions inherent to the MRI data while promoting robustness to spatial variations. All data processing steps were implemented using the MONAI (version 1.4.0).

2.3 Model Development for Task I

We used a base-size Vision Transformer (ViT) model with a 2D image embedding layer as the encoder. The encoder was implemented using the timm library (version 1.0.15), and was initialized with pre-trained weights from the base-size Vision Transformer (ViT-B/16). Key encoder parameters include an embedding dimension of 768, a patch size of 16, LayerNorm for pre-normalization, and a dropout rate of 0.1 during training. A schematic of the model architecture is shown in Fig. 1.

Task I involves multi-category prediction across 7 distinct categories, each with 3 possible levels of image quality. Consequently, the final fully connected (FC) layer was configured with $7 \times 3 = 21$ output nodes. After computing the loss, the 21-dimensional output vector is reshaped into a 7×3 matrix. A Softmax function is applied independently to each of the 7 category rows, followed by the calculation of the cross-entropy loss.

During the training stage, the model was trained using pairs of individual slices and their corresponding volume-level labels. During inference, all slices from a given MR volume are predicted individually, generating a prediction tensor of shape n_slices $\times$ 7 $\times$ 3. To aggregate these slice-level predictions into a single volume-level prediction, the maximum value across all slices is taken for each category and quality level combination, resulting in a 7×3 matrix. The final quality level for each category is then determined by applying the argmax function along the quality level dimension (axis $= 1$) of this aggregated matrix.

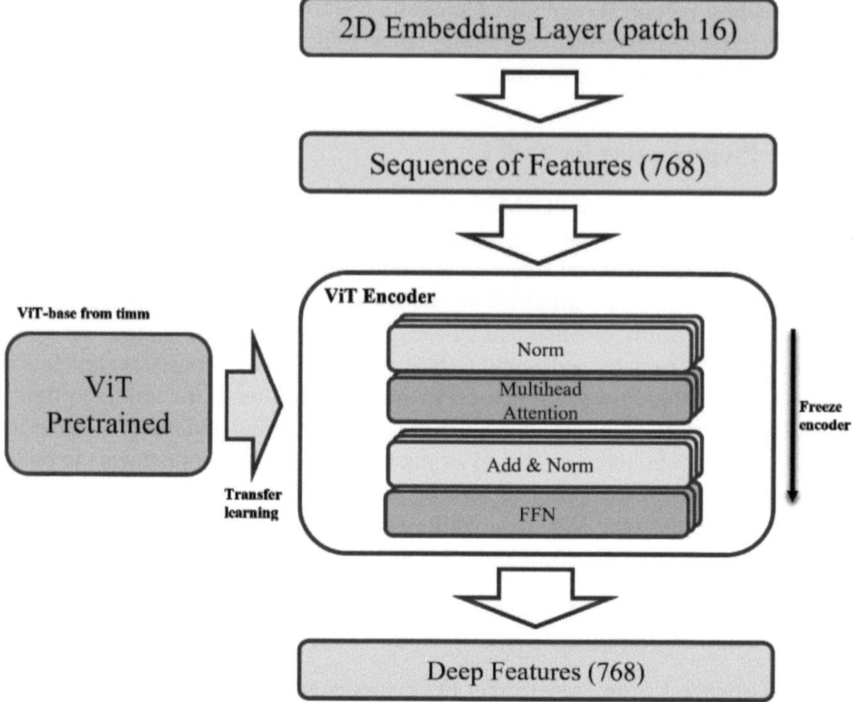

Fig. 1. Schematic of the model architecture. The base structure is the ViT-base model with 768-dimensional embedding features, pre-trained weights from timm. During training, the parameters of the ViT encoder were frozen, and only the final fully connected (FC) layer was trained.

Regarding hyperparameters, we used a batch size of 256 and trained for 200 epochs. A cosine annealing learning rate scheduler was employed, decaying the initial learning rate of 0.001 to 0. We performed cross-validation on the LISA training set. For the final submitted model, we selected the checkpoint that achieved the highest average AUROC on our internal validation fold. All development was done using PyTorch (version 2.7.0) and the Transformers library (version 4.46.2). To reduce GPU memory consumption, we used mixed-precision training and inference via torch.autocast with float16 precision. Training was conducted on two NVIDIA RTX 3090 GPUs.

2.4 Task II Segmentation

Task II involves two distinct segmentation sub-tasks: the first is the segmentation of bilateral hippocampi (3 classes), and the second is the segmentation of basal ganglia (6 classes). We trained separate models for each sub-task. The entire data processing and model development pipeline remained consistent for both.

2.5 Image Preprocessing for Task II

The original MR images were first normalized using Z-normalization. To reduce GPU memory consumption and increase the effective sample size, we partitioned the 3D MR images into patches of size 128x128x128. This patchification was performed using the GridSampler from the torchio library (version 0.20.4), with an overlap set to 64x64x64. During training, only patches containing positive labels (foreground) were included in the training queue. However, for the validation set, all patches were retained to enable whole-volume evaluation. For online data augmentation, we used both spatial and intensity transformations. Spatial augmentations included random flipping along the axial, coronal, and sagittal planes, as well as random affine transformations. Intensity augmentations consisted of random blur and random gamma adjustments. All augmentations were implemented using the MONAI framework.

2.6 Model Development for Task II

We used the UNETR architecture provided by MONAI as our segmentation model. The model was initialized with pre-trained weights from SAM-Med 3D. Although SAM-Med 3D is designed for instance segmentation, we extracted its encoder parameters and transferred them to the ViT encoder within the UNETR network. According to the SAM-Med 3D configuration, the corresponding UNETR network was set up with the following parameters: image_size $= 128$, patch_size $= 16$, hidden_size $= 768$, mlp_dim $= 3072$, num_heads $= 12$. The model uses Instance Normalization (IN) layers and a dropout rate of 0.1 during training. A schematic diagram of the model pipeline is provided in Fig. 2.

Training was conducted in two stages:

- Stage 1: The ViT encoder of the UNETR model was frozen, and only the decoder (the U-Net part) was trained.
- Stage 2: LoRA was applied to the query, key, value (qkv) projections and linear layers within the encoder, with a rank r $= 4$ and alpha $\alpha = 16$, followed by fine-tuning the entire model with these adapters.

The loss function was a combined Cross-Entropy and Dice Loss. For training hyper-parameters, we used a batch size of 16 on two RTX 3090 GPUs, optimized with AdamW, and trained for 80 epochs, including a 5-epoch warm-up phase. Mixed precision (float16) was employed during both training and inference. Model development we used 5-fold cross-validation. For the final submission, the model fold achieving the highest mean Dice (mDICE) on the internal validation set was selected.

2.7 Statistical Analysis

For Task I (Quality Assessment), we used the F1, F2, accuracy, and averaged accuracy metrics. For Task II (Segmentation), the evaluation was based on Dice score, Hausdorff Distance (HD), Average Symmetric Surface Distance (ASSD), and Relative Volume Error (RVE). The internal validation results were calculated on a subset of the challenge training dataset during model development, while the validation and testing statistics

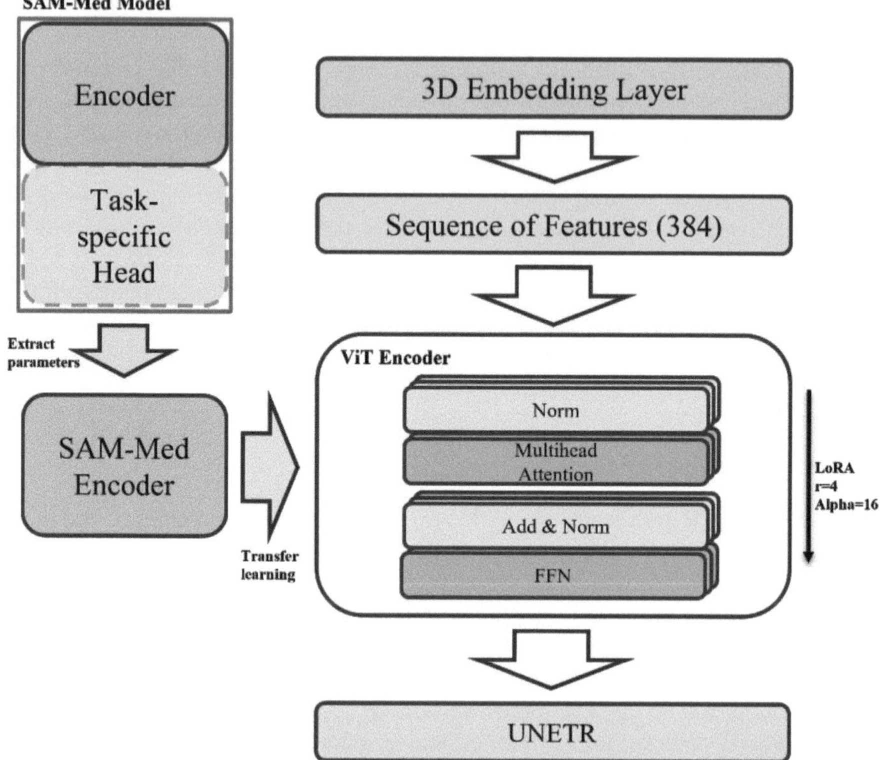

Fig. 2. We used UNETR as the segmentation model. The encoder of the SAM-Med 3D network was used to provide pre-trained parameters for the ViT component of UNETR (embedding dimension 768). LoRA with r = 4 and alpha = 16 was applied during the fine-tuning stage.

were derived from the official LISA 2025 Challenge evaluation. Classification statistical analyses were performed using scikit-learn (version 1.5.2).

3 Results

3.1 Task I Quality Assessment Results

The performance for each quality category, based on the validation set derived from the challenge model development dataset, is summarized in Table 1.

The performance on the validation and testing sets, based on the official online evaluation, is summarized in Table 2.

Table 1. Internal (offline) validation results for Task I (Image Quality Assessment)

	Precision	Recall	F1_Score	F2_Score	Accuracy
Noise	0.5588	0.6	0.5781	0.5909	0.8721
Zipper	0.5536	0.4905	0.5113	0.4969	0.8721
Positioning	0.3101	0.3333	0.3213	0.3284	0.9302
Banding	0.314	0.3333	0.3234	0.3293	0.9419
Motion	0.2674	0.3333	0.2968	0.3177	0.8023
Contrast	0.3968	0.3457	0.2953	0.3211	0.686
Distortion	0.2946	0.3333	0.3128	0.3248	0.8837
Macro Average	0.3851	0.3956	0.377	0.387	0.8555

Table 2. Official validation and testing results for Task I (Image Quality Assessment)

	Precision	Recall	F1	F2	Accuracy	Average
Validation stage	0.786	0.861	0.821	0.844	0.861	0.834
Testing stage	0.743	0.831	0.781	0.809	0.831	0.799

3.2 Task II Segmentation Results

The performance on the validation and testing sets is summarized in Table 3.

Table 3. Official validation and testing results for Task II (Segmentation)

	Stage	DSC	HD	HD95	ASSD	RVE
Bilateral Hippocampi	Validation	0.69 ± 0.17	6.32 ± 8.52	2.28 ± 0.96	0.73 ± 0.53	0.14 ± 0.08
	Testing	0.58 ± 0.19	4.61 ± 1.96	3.15 ± 1.54	1.06 ± 0.62	0.18 ± 0.14
Basal Ganglia	Validation	0.86 ± 0.05	3.10 ± 0.78	1.65 ± 0.77	0.46 ± 0.23	0.07 ± 0.04
	Testing	0.81 ± 0.10	3.54 ± 1.27	1.95 ± 1.01	0.67 ± 0.40	0.17 ± 0.18

Figure 3 shows representative axial slices showing segmentation results on the internal validation set. The model predictions tend to favor higher precision, though false negatives (missed regions) are still observed.

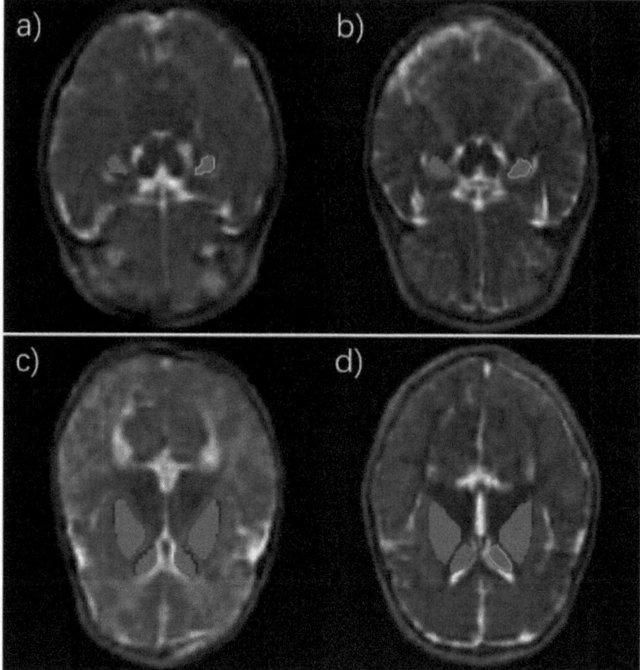

Fig. 3. Task II segmentation results on the internal validation set. Yellow indicates correctly segmented regions, green highlights false negatives (missed by the model), and red denotes false positives. (a) and (b) show hippocampal segmentation examples with Dice scores of 0.690 (lowest) and 0.833 (highest) within the validation set, respectively. (c) and (d) show basal ganglia segmentation with Dice scores of 0.643 and 0.897, respectively.

4 Discussion

For task I, during the development phase, we compared Low-Rank Adaptation (LoRA) fine-tuning (with rank $r = 4$, alpha $= 16$) against the strategy of freezing the entire encoder and only training the final FC layer. We found that incorporating LoRA did not yield superior performance on our internal validation set compared to simply freezing all encoder parameters. Furthermore, given our slice-wise training approach, aggregating the 2D slice predictions into a consolidated 3D volume prediction was necessary during inference. We experimented with two aggregation methods: taking the mean and taking the maximum prediction value across slices for each class/quality level. Our evaluation on the internal validation set indicated that using the maximum value for aggregation slightly outperformed using the mean value.

For task II, based on our evaluation on the internal validation set, the incorporation of LoRA fine-tuning in the second stage yielded slightly superior Dice scores compared to training without it. The segmentation performance appears to be correlated with the contrast and size of the target structures. The second sub-task (Basal Ganglia) demonstrated significantly higher segmentation accuracy across all metrics compared to the first sub-task (Bilateral Hippocampi), which is likely attributable to the larger volume and more distinct contrast of the basal ganglia nuclei in the MR images.

5 Conclusion

Our training pipeline is designed to be simple and straightforward. For this challenge, we primarily used ViT architectures as the core component, adapting them with different task-specific heads for both image quality assessment (classification) and segmentation.

For Task I (Image Quality Assessment), we converted 3D volumes into 2D slices for training. The final volume-level quality assessment result was obtained by taking the maximum prediction value across all slices for each category and quality level. We employed a classification head with $7 \times 3 = 21$ output nodes to handle the multi-category and multi-level task. During training, we used a ViT-base model pre-trained on ImageNet (from the timm library) and froze the entire 2D ViT encoder, training only the final classification layer.

For Task II (Segmentation), we trained on 3D patches. We extracted the encoder from SAM-Med 3D to initialize our model. Our segmentation network was based on the UNETR architecture. The training process consisted of two stages: the first stage involved freezing the ViT encoder and training only the U-Net decoder, while the second stage applied LoRA with a rank $r = 4$ and alpha $\alpha = 16$ to fine-tune the encoder, which resulted in a slight performance improvement.

6 Limitations and Future Work

A limitation of our current approach is the source of our pre-trained weights. For Task I, we used a 2D ViT model pre-trained on ImageNet, a natural image dataset. For Task II, we used weights from SAM-Med 3D, which, although medical, was pre-trained for an instance segmentation task. This represents a form of post-pre-training adaptation for our specific downstream tasks.

Future work could focus on leveraging pure, large-scale pre-training methods specifically designed for medical imaging. Using models pre-trained with frameworks like medical-domain CLIP (Contrastive Language-Image Pre-training) or Masked Autoencoders (MAE) on extensive volumetric medical data could potentially provide more robust and transferable feature representations, leading to further performance improvements in both quality assessment and segmentation tasks. Exploring end-to-end training strategies without freezing the encoder could also be investigated with more extensive computational resources.

Disclosure of Interests. The authors have no competing interests.

References

1. He, K., Zhang, X., Ren, S., Sun, J.: Deep residual learning for image recognition. In: Proceedings of the IEEE Conference on Computer Vision and Pattern Recognition, pp. 770–778 (2016)
2. Ronneberger, O., Fischer, P., Brox, T.: U-net: convolutional networks for biomedical image segmentation. In: International Conference on Medical Image Computing and Computer-Assisted Intervention, pp. 234–241. Springer, Cham (2015)

3. Dosovitskiy, A., et al.: An image is worth 16×16 words: transformers for image recognition at scale. arXiv preprint arXiv:2010.11929 (2020)
4. Radford, A., et al.: Learning transferable visual models from natural language supervision. In: International Conference on Machine Learning, pp. 8748–8763. PMLR (2021)
5. Arnab, A., Dehghani, M., Heigold, G., Sun, C., Lučić, M., Schmid, C.: Vivit: a video vision transformer. In: Proceedings of the IEEE/CVF International Conference on Computer Vision, pp. 6836–6846 (2021)
6. Hatamizadeh, A., et al.: Unetr: transformers for 3D medical image segmentation. In: Proceedings of the IEEE/CVF Winter Conference on Applications of Computer Vision, pp. 574–584 (2022)
7. Ma, J., He, Y., Li, F., et al.: Segment anything in medical images. Nat. Commun. **15**, 654 (2024). https://doi.org/10.1038/s41467-024-44824-z
8. Wang, H., et al.: SAM-Med3D: towards general-purpose segmentation models for volumetric medical images. In: European Conference on Computer Vision, pp. 51–67. Springer Nature Switzerland, Cham (2024)
9. Lepore, N., Linguraru, M.G.: Low Field Pediatric Brain Magnetic Resonance Image Segmentation and Quality Assurance: First MICCAI Challenge, LISA 2024, Held in Conjunction with MICCAI 2024, Marrakesh, Morocco, October 10, 2024, Proceedings, p. 77. Springer Nature (2025)

Automatic Quality Assurance and Subcortical Brain Segmentation in Pediatric Ultra-Low-Field MRI: Exploring Ordinal Learning and Foundation Model Adaptation

Raquel González López[1]([envelope]), Maria Chiara Fiorentino[2], Gerard Martí-Juan[1], Oscar Camara[1], and Miguel A. González Ballester[1,3]

[1] BCN Medtech, Department of Engineering, Universitat Pompeu Fabra,
Barcelona, Spain
`{raquel.gonzalez,gerard.marti,oscar.camara,ma.gonzalez}@upf.edu`
[2] Department of Information Engineering, Università Politecnica Delle Marche,
Ancona, Italy
`m.c.fiorentino@staff.univpm.it`
[3] ICREA, Barcelona, Spain

Abstract. Ultra-low-field (uLF) MRI systems offer portable and affordable neuroimaging solutions for pediatric patients and are valuable in resource-limited settings. However, such systems are susceptible to poor image quality, artifacts, and low contrast, making brain segmentation difficult. This study addresses two critical challenges in uLF MRI: automated quality assessment (QA) and anatomical structure segmentation. For QA, we propose a multi-label approach that incorporates the ordinal nature of artifact severity through an ordinal loss and models artifact co-occurrence patterns using Bayesian Networks. The approach is enhanced through aggressive synthetic data augmentation and ensemble learning, achieving a composite accuracy score of 0.84 across seven artifact categories. For segmentation, we benchmark a task-specific model (nnU-net) against a foundation model (SAM-Med3D) on the delineation of challenging subcortical structures. While nnU-Net, trained from scratch, achieved mean Dice score of 0.72 for hippocampi and 0.86 for basal ganglia, we demonstrate that lightweight fine-tuning of SAM-Med3D yields comparable results with a mean Dice score of 0.70 in hippocampi segmentation, despite domain shift. These results underscore the promise of foundation models for medical imaging in low-resource contexts, while highlighting the importance of domain adaptation. Overall, our pipeline represents a step forward in robust, automated QA and segmentation in

---

The original version of the chapter has been revised. An acknowledgments section has been added to Chapter 11. A correction to this chapter can be found at
https://doi.org/10.1007/978-3-032-14417-1_12

Supplementary Information The online version contains supplementary material available at https://doi.org/10.1007/978-3-032-14417-1_11.

uLF MRI for pediatric use. We release the code at https://github.com/reitxel/LISA2025TeamUPF.

Keywords: Low-field pediatric MRI · Quality assurance · Image Segmentation

1   Introduction

Magnetic resonance imaging (MRI) is a fundamental tool for studying the developing brain, offering non-invasive and high-resolution insights into fetal and neonatal neuroanatomy [19]. However, conventional high-field MRI systems (1–5-3T), while clinically powerful, are often prohibitively expensive and require dedicated infrastructure such as shielded rooms and highly trained operators. These constraints make access to MRI highly unequal worldwide, especially in low- and middle-income countries, where the majority of the population lacks access to this technology [10]. To overcome these limitations, ultra-low-field (uLF) MRI systems, such as the Hyperfine Swoop (64mT), have been proposed as affordable and portable alternatives. These systems can be used on the bedside without the need for specialised facilities, making them particularly suitable for neonatal care units and remote or underserved areas. However, despite their practical advantages, uLF MRI remains susceptible to a wide range of artifacts that can significantly compromise image quality. In the perinatal setting, frequent fetal or neonatal motion introduces motion blur and ghosting; signal inhomogeneity and low tissue contrast further degrade the visual quality of the images. These limitations are exacerbated by the inherently reduced spatial resolution of low-field systems, making the accurate visualization and analysis of fine neuroanatomical structures especially challenging [2]. In this context, machine learning (ML) and deep learning (DL) models have shown strong potential in addressing some of these limitations, also thanks to the community's efforts in releasing data through structured initiatives such as Challenges. Notably, the *Low-field Pediatric Brain Magnetic Resonance Image Segmentation and Quality Assurance (LISA) Challenge 2024* [8], held in conjunction with MICCAI, focused on two key tasks essential for the clinical adoption of uLF MRI: (1) automatic image quality assurance (QA) and (2) hippocampi segmentations, which are central to memory and cognitive function and are often implicated in abnormal neurodevelopment.

For QA, the top-performing approaches included a CNN encoder combined with Mamba blocks [23]—lightweight state-space models that effectively capture global contextual information without the computational burden of traditional Transformers—and a DenseNet backbone enhanced with targeted data augmentations [6,16,17]. However, most existing methods treat artifact categories independently and ignore the ordinal nature of severity levels. To address these limitations, we incorporate an ordinal loss to reflect severity progression and model artifact co-occurrence patterns through Bayesian networks, enabling the system to exploit shared dependencies among artifact types. To further mitigate class imbalance, we introduce a targeted synthetic augmentation strategy to generate realistic artifacts across severity levels.

For the segmentation task, most solutions adopted encoder-decoder architectures based on U-net or its variants (V-net, nnUnet) [6,17,22]. Some approaches incorporated prior knowledge or anatomical constraints to guide the learning process [16], while others employed adversarial learning between raw low-field images and their frequency-filtered counterparts [13]. However, the potential of foundation models such as the Segment Anything Model (SAM) Med3D [21], remained largely unexplored, particularly their performance on challenging tasks involving anatomical complexity-like hippocampal segmentation-as well as noise and contrast issues that are amplified in uLF MRI. In this work, we evaluate the adaptability of SAM-Med3D through lightweight fine-tuning and benchmark its performance against nnU-Net, a widely adopted task-specific convolutional neural network in the field. In addition to hippocampal segmentation, we also develop a dedicated pipeline for basal ganglia, a set of nuclei involved in motor control, behavior regulation, and executive functioning, further contributing to automated neuroanatomical analysis in the uLF pediatric MRI setting.

2   Methods

2.1   Datasets

This study draws primarily on the datasets provided by the LISA 2025 Challenge. Both datasets consist of uLF 0.064T scans captured by the Hyperfine SWOOP portable scanner across different sites.

For *QA*, the dataset comprises 532 T2-weighted spin echo scans from 244 different subjects, acquired with TR = 1.5 s, TE = 5 ms, TI = 400 ms. Each subject has up to three scans, captured at different orthogonal planes. Data acquisition was performed at three different institutions: University of Cape Town, (Cape Town, South Africa), Kawempe National Referral Hospital, (Makerere University, Kampala, Uganda), and Aga Khan University Hospital, (Karachi, Pakistan). Each scan was evaluated across seven artifact domains: noise, zipper, positioning, banding, motion, contrast and distortion, using a three-point scale: 0 (no visible artifact), 1 (artifact present with minimal impact on neural structure visualization), and 2 (severe artifact significantly impairing neural structure differentiation).

For *anatomical segmentation*, the dataset released consists of 79 uLF T2-weighted MRI scans collected across multiple institutions, including Kawempe National Referral Hospital (Makere University, Uganda), CUBIC (University of Cape Town, South Africa), the Warren Alpert Medical School of Brown University, and the Advanced Baby Imaging Lab at Rhode Island Hospital (Providence, RI, USA). All scans were acquired by experienced MRI technicians at each site. Although each low-field image has a corresponding high-field counterpart, only the low-field scans were released. Expert-reviewed manual segmentations of the bilateral hippocampi, ventricles and basal ganglia are provided to support training and evaluation of segmentation models.

2.2   Task 1: Quality Assurance (QA)

We address the multi-label MRI quality control challenge as a structured multitask learning problem with seven distinct artifact categories: Noise, Zipper, Positioning, Banding, Motion, Contrast, and Distortion. Each category can have 3 labels: 0 (normal), 1 (mild), and 2 (severe). The provided dataset exhibits severe class imbalance, with approximately 80% of the samples belonging to class 0, presenting significant challenges for model training, validation and evaluation.

Data Augmentation. To improve robustness to the wide variability and imbalanced data artifacts, we adopted a two-stage augmentation strategy, applied both offline and online (for DL models). The offline data augmentation strategy followed the synthetic artifact generation pipeline proposed in [16], adapting the parameters to our training set. In two of the classes, we changed the original approach: for Zipper, we implemented a band-based approach that generates narrow, randomly positioned vertical or horizontal bands with variable thickness, incorporating intensity variation and strong noise within bands. For Distortion, we combined the elastic deformation from the original approach with complementary artifacts, including bias field inhomogeneity, ghosting, and spike artifacts. Full parameters of the data augmentation procedure for each artifact can be found in Supplementary File S1. Online data augmentation was applied in batch consisting of random shift and scaling in intensity, random rotations, and affine transformations. To ensure realism, augmentation parameters were tuned through iterative visual comparisons with real low-field artifacts, preserving label semantics and motivating deviations from the parameterizations in [16].

Models. We compared two different approaches: a classical ML approach, using handcrafted features, and a DL based one.

For the classical ML approach, we segmented the brain of the images using BET [5] with a conservative threshold of $f = 0.2$ to ensure full brain coverage. From each image, we extracted 29 handcrafted features, including basic image quality metrics (e.g. signal-to-noise ratio, signal range, texture homogeneity), histogram entropy, frequency domain characteristics and texture descriptors. Tissue segmentation was estimated through intensity thresholding. Full feature definitions are available in Supplementary File S2. We trained a Random Forest (100 trees) model with those features, evaluating it using ten-fold cross-validation.

For the DL approach, we trained separate models for each task, without direct weight sharing or multitask learning. Each model uses a 3D Densely connected convolutional network (DenseNet) [4] backbone, with three different losses: Cross Entropy loss, Focal loss [9] and Earth Mover's Distance (EMD) loss [3], also known as Ranked Probability Score (RPS).

The Focal loss is defined as:

$$L_{Focal}(p_{t,i}) = -\alpha_{y_i}(1 - p_{t,i})^\gamma \log(p_{t,i}), \tag{1}$$

where $p_{t,i}$ represents the predicted probability for class i, α_{y_i} a class-specific weighting, and $\gamma \geq 0$ is the focusing parameter that reduces the loss for well-classified examples, putting more emphasis on misclassified, harder samples. This loss has been shown to work well for unbalanced problems [9].

We use the EMD loss to evaluate the problem as an ordinal classification problem. The underlying idea is that the labels in the problem have an inherent order (the artifact severity), and a misclassification from 2 to 0, for example, should be more penalized than from 1 to 2. The EMD loss is defined as:

$$L_{EMD}(y, \hat{y}) = \sum_{i=1}^{C} \left(\mathrm{CDF}y(i) - \mathrm{CDF}\hat{y}(i) \right)^2, \tag{2}$$

where $\mathrm{CDF}y$ and $\mathrm{CDF}\hat{y}$ are the cumulative distribution functions of the true and predicted distributions, respectively, and C is the number of classes.

All models have been trained over 100 epochs with a learning rate (lr) of $4e - 3$ using Adam optimizer [7] and cosine annealing schedule. We used early stopping (if the validation loss did not improve over 15 epochs) to reduce overfitting. Images were cropped and resized to (150,150,150) before input to facilitate training. The evaluation of the challenge consists of micro, macro and weighted average for all 7 tasks, over 5 standard metrics: Accuracy, F1-score, F2-score, Precision and Recall. The final challenge score is computed as the arithmetic mean of these weighted metrics.

Bayesian Networks. The occurrence of artifacts on uLF MRI images is not random: some of the artifacts are caused by similar underlying physical phenomena, patient-related factors, or hardware limitations, leading to strong co-occurrence patterns and conditional dependencies between different distortion types. We decided to create a Bayesian network (BN) model [1] to incorporate this information into our prediction. We define the BN as a fully connected Directed Acyclic Graph (DAG), with edges (i, j) for all $i < j$ in topological order, where i and j represent the artifact classes. Structure learning employs Maximum Likelihood Estimation on training fold data.

During inference, we perform hard evidence propagation using Variable Elimination. For each distortion category i, evidence is collected from predictions of other categories $j \neq i$ where $\max(P_j) > 0.6$. The final probability adjustment combines deep learning and BN predictions:

$$P_{\mathrm{adj}}(X_i) = (1 - \alpha) \cdot P_{\mathrm{DL}}(X_i) + \alpha \cdot P_{\mathrm{BN}}(X_i | \mathrm{Evidence}), \tag{3}$$

with $\alpha = 0.2$ determined through validation experiments.

Ensemble and Evaluation. We implemented an ensemble approach to combine multiple models for each of the seven tasks. For each artifact category, we train several models with different hyperameters, data augmentation techniques, and losses, and then combine their predictions to obtain robust probability estimates. The resulting predictions serve as input to the Bayesian Network stage, obtaining the final prediction. This task-specific ensemble strategy reduces false positives and improves generalization.

We have tested different combinations of losses, augmentations, and ensembling models, finding the best thresholds and hyperparameters of the Bayesian

Network. All tests have been done over a separation validation set, with 80/20 separation. Models have been built using Pytorch and trained on a shared High Performance Computing cluster, with a Nvidia Tesla T4 with 16 GB of VRAM.

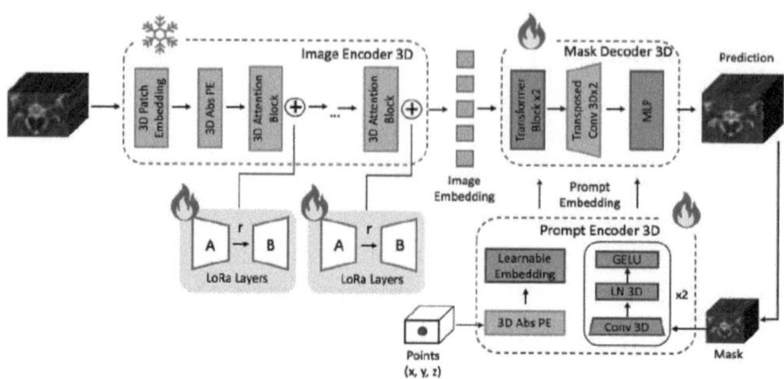

Fig. 1. Overview of the SAM-Med3D architecture employed, including the Parameter Efficient Fine-Tuning (PEFT) strategy based on Low-Rank Adaptation (LoRA), integrated in the image encoder.

2.3   Task 2 + Task 3: Hippocampal and Basal Ganglia Segmentation

Hippocampal Segmentation. As introduced in Sect. 1, we explored the feasibility of using promptable foundation models for hippocampal segmentation in uLF MRI, focusing on SAM Med3D [21]. This approach leverages the transferability of large-scale pre-trained representations, offering a potentially powerful alternative to traditional segmentation pipelines that rely on handcrafted architectures or extensive task-specific supervision. To evaluate its suitability for uLF MRI, we systematically compared it against nnU-Net, a widely adopted benchmark in the LISA Challenge 2024 and a strong state-of-the-art baseline for 3D medical image segmentation. Fig. 1 illustrates the architecture used. To adapt the SAM-Med3D model to the uLF MRI domain without updating the full set of model parameters (avoiding overfitting), we adopted a Parameter-Efficient Fine-Tuning (PEFT) strategy based on Low-Rank Adaptation (LoRA). In the standard multi-head self-attention (MHSA) mechanism, the query, key, and value matrices are computed as follows:

$$Q, K, V = XW_Q, XW_K, XW_V, \quad \text{with} \quad W_Q, W_K, W_V \in \mathbb{R}^{d \times d}$$

In our LORA-based approach, the original weight matrices are kept frozen, and a trainable low-rank update is introduced:

$$W \leftarrow W + \alpha \cdot BA$$

where $A \in \mathbb{R}^{d \times r}$ and $B \in \mathbb{R}^{r \times d}$ are learnable matrices with rank $r \ll d$, and α is a scaling factor.

Applied to the attention projections, this gives:

$$XW_i + \alpha X A_i B_i, \quad \text{for } i \in \{Q, K, V\}$$

This strategy significantly reduces the number of trainable parameters and memory footprint, while retaining the strong priors of the pre-trained model. We applied LoRA specifically to the qkv projection layers of each attention block in the encoder.

Experimental Setup. Both models were trained using a five-fold cross-validation strategy. Final segmentation maps were obtained by averaging the predicted probabilities across the folds, followed by thresholding to obtain the final binary segmentation.

Regarding SAM-Med3D, the model was initialized with pre-trained SAM-Med3D weights provided in [21]. Input volumes were processed to $128 \times 128 \times 128$ voxels with canonical orientation, Z-score normalization and random flip augmentation. Training was conducted for 100 epochs using AdamW optimizer ($lr = 8 \times 10^{-4}$, weight decay $= 0.1$) with batch size $= 1$ and r $= 4$. We employed a DiceCELoss with sigmoid activation and mixed precision training. Interactive segmentation used random point sampling with 1–20 clicks per iteration.

The nnU-Net was trained using the 3D full-resolution pipeline, with a patch size of $160 \times 224 \times 192$, reflecting the median image dimension of the dataset. A batch size of 1 was used due to GPU memory constraints. Preprocessing included Z-score normalization applied across the entire image. The architecture was based on a six-stage Residual Encoder UNet, using 3D convolutions, instance normalization, and LeakyReLU activation combined both spatial and intensity-based transformations. Spatial augmentation included random rotations and scaling. Intensity augmentations involved the addition of Gaussian noise and blur, random adjustments of brightness and contrast, gamma correction, and simulated low-resolution inputs. Additional enhancements included histogram equalization and frequency-domain perturbations. Random mirroring was performed along spatial axes to improve robustness.

Basal Ganglia Segmentation. For the segmentation of the basal ganglia, we targeted four anatomical structures: the left and right caudate nuclei and the left and right lentiform nuclei. The same nnU-Net configuration and training protocol described in the Experimental Setup section was applied to this task.

Evaluation Metrics. To quantitatively assess the segmentation performance, we employed the five official metrics of the LISA Challenge 2025 for all the structures, then averaged: Dice Similarity Coefficient (DSC), Hausdorff Distance (HD), 95th percentile HD (HD95), Average Symmetric Surface Distance (ASSD), and Relative Volume Error (RVE).

3   Results

3.1   Task 1: QA

After various experiments and parameter exploration, we have evaluated 5 different models: one classic ML model, using Random Forest, and 4 deep learning based models, all of them using both offline and online augmentations, with a DenseNet architecture and different loss functions: Cross Entropy loss, Focal loss, EMD loss, and a combined Focal + EMD loss, with $\lambda_{focal} = 0.4$ and $\lambda_{EMD} = 0.6$, chosen experimentally. The final ensemble is done combining the X, Y and Z models, as this was the combination scoring the best.

Table 1 shows the results of the QA task, across various models and ensembles, for both the validation set and the online validation benchmark of the challenge.

3.2   Task 2 + Task 3: Anatomical Segmentations

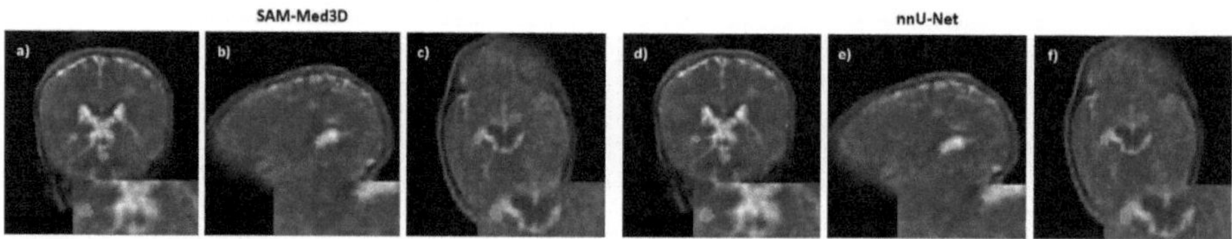

Fig. 2. Example segmentations of the left (red) and right (green) hippocampi overlaid on uLF MRI images. Predictions from SAM-Med3D$_{10}$ are shown in (a) coronal, (b) sagittal, and (c) axial views, while nnU-Net predictions are displayed in (d) coronal, (e) sagittal, and (f) axial views. (Color figure online)

Table 2 presents the evaluation results for both hippocampal and basal ganglia segmentation. For the hippocampi, we compare nnU-Net with SAM-Med3D using single (SAM-Med3D$_1$)- and 10-point (SAM-Med3D$_{10}$) prompt configurations derived from nnU-Net segmentations, which serve as one of the gold standards in the field. While segmentation was consistently more challenging for the left hippocampus, nnU-Net outperformed SAM-Med3D on all metrics except RVE, where SAM-Med3D$_{10}$ achieved a slightly lower value (0.14 vs. 0.17). As shown in Fig. 2, nnU-Net predictions appear more spatially coherent, while SAM-Med3D results tend to be less smooth. Increasing prompt density improved SAM-Med3D performance across all metrics.

For the basal ganglia, nnU-Net achieved high DSC in all structures, with slightly better performance on the right side (DSC = 0.87). Distance-based metrics and RVE remained low throughout, indicating accurate boundary delineation and volume estimation.

Table 1. Performance comparison across models on training and evaluation datasets. Acc.: Accuracy. Prec.: Precision. RF: Random Forest. CE: Cross Entropy loss. EMD: Earth Mover's Distance loss. BN: Bayesian Network.

Metric	Model					
	RF	CE loss	EMD loss	Focal loss	EMD+Focal	Ensemble+BN
Validation set						
Acc. (Micro)	0.619	0.762	0.867	0.838	0.848	**0.874**
Acc. (Macro)	**0.396**	0.290	0.330	0.319	0.322	0.329
Acc. (Weighted)	0.619	0.762	0.867	0.838	0.848	**0.874**
F1 (Micro)	0.619	0.762	0.867	0.838	0.848	**0.874**
F1 (Macro)	**0.340**	0.288	0.310	0.304	0.306	**0.311**
F1 (Weighted)	0.683	0.758	0.814	0.799	0.804	**0.825**
F2 (Micro)	0.619	0.762	0.867	0.838	0.848	**0.874**
F2 (Macro)	**0.356**	0.289	0.321	0.313	0.316	0.322
F2 (Weighted)	0.636	0.760	0.845	0.822	0.830	**0.853**
Prec. (Micro)	0.619	0.762	0.867	0.838	0.848	**0.874**
Prec. (Macro)	**0.359**	0.287	0.292	0.290	0.291	0.294
Prec. (Weighted)	**0.802**	0.754	0.767	0.763	0.765	0.781
Recall (Micro)	0.619	0.762	0.867	0.838	0.848	**0.874**
Recall (Macro)	**0.396**	0.290	0.330	0.319	0.322	0.329
Recall (Weighted)	0.619	0.762	0.867	0.838	0.848	**0.874**
Composite	0.672	0.759	0.832	0.812	0.819	**0.841**
Online validation benchmark						
Acc. (weighted)	–	0.796	**0.871**	0.85	0.857	**0.871**
F1 (weighted)	–	0.801	**0.833**	0.816	0.825	**0.833**
F2 (weighted)	–	0.797	**0.854**	0.836	0.843	**0.854**
Prec. (weighted)	–	0.813	**0.822**	0.787	0.799	**0.822**
Recall (weighted)	–	0.796	**0.871**	0.85	0.857	**0.871**
Composite score	–	0.801	**0.850**	0.828	0.836	**0.850**

Table 2. Segmentation performance for hippocampi (SAM Med3D and nnU-Net) and basal ganglia (nnU-Net only) on validation set. Metrics reported as mean (standard deviation). Best performances in bold.

Region	Model	DSC↑	HD ↓	HD95 ↓	ASSD ↓	RVE ↓
Hipp L	SAM Med3D$_1$	0.64 (0.21)	9.14 (15.87)	2.69 (1.78)	1.03 (1.14)	0.21 (0.16)
	SAM Med3D$_{10}$	0.66 (0.21)	9.75 (16.34)	2.64 (2.01)	1.00 (1.33)	**0.16 (0.14)**
	nnU-Net	**0.69 (0.23)**	**8.50 (16.15)**	**2.45 (2.32)**	**0.95 (1.46)**	0.20 (0.13)
Hipp R	SAM Med3D$_1$	0.69 (0.14)	4.20 (1.13)	2.40 (1.08)	0.73 (0.40)	0.21 (0.14)
	SAM Med3D$_{10}$	0.73 (0.16)	4.08 (0.97)	1.95 (0.81)	0.58 (0.42)	**0.12 (0.09)**
	nnU-Net	**0.75 (0.15)**	**3.36 (0.90)**	**1.60 (0.92)**	**0.51 (0.42)**	0.13 (0.08)
Hipp Avg	SAM Med3D$_1$	0.67 (0.17)	6.67 (7.94)	2.55 (1.40)	0.88 (0.76)	0.21 (0.11)
	SAM Med3D$_{10}$	0.70 (0.18)	6.91 (8.18)	2.30 (1.38)	0.79 (0.87)	**0.14 (0.09)**
	nnU-Net	**0.72 (0.19)**	**5.93 (8.13)**	**2.02 (1.59)**	**0.73 (0.93)**	0.17 (0.09)
Caudate L		0.84 (0.06)	3.69 (1.52)	1.63 (0.62)	0.42 (0.20)	0.10 (0.07)
Caudate R		0.87 (0.06)	3.17 (1.35)	1.45 (0.78)	0.36 (0.24)	**0.06 (0.05)**
Lentiform L	nnU-Net	0.86 (0.05)	3.02 (0.98)	1.89 (0.96)	0.58 (0.27)	0.07 (0.08)
Lentiform R		0.87 (0.05)	2.68 (0.78)	1.59 (0.66)	0.51 (0.25)	0.08 (0.07)
BG Avg		**0.86 (0.05)**	**3.14 (0.81)**	**1.64 (0.68)**	**0.47 (0.21)**	0.08 (0.04)

4 Discussion

This study presents a novel strategy to address two key challenges in uLF MRI, within the context of LISA Challenge 2025: automated QA and anatomical segmentation. For QA task, we proposed an ensemble of DL methods based on DenseNet backbone, incorporating both classification and ordinal regression losses. The models were trained with extensive data augmentation to enhance robustness to variability in artifact appearance and to mitigate the effects of data imbalance. To further enhance performance, a Bayesian network was integrated to model artifact co-occurrence across tasks. For hippocampal segmentation, we investigated whether the domain knowledge encoded in foundation models-originally trained in different imaging contexts-can be effectively leveraged in the uLF MRI domain. Specifically, we adapted SAM-Med3D using LoRA and compared its performance to the established nnU-Net framework.

Our approach for uLF QA demonstrates the effectiveness of combining DL architectures with different learning objectives to address the challenge. Ordinal loss (EMD) proved particularly effective across tasks, achieving a composite score of 0.85 on online validation by appropriately penalizing larger prediction errors (e.g., classifying severe artifacts as normal) and facilitating stable training in the presence of class imbalance. Our Bayesian network approach showed marginal improvements in our validation set (mainly in removing false positives), but the same performance as EMD loss in the online set. RF showed bad results, although marginally better micro metrics. Our experiments revealed that enhancing data augmentation approaches to generate synthetic artifacts closely resembling real distortions was essential for achieving competitive performance. This highlights that severe data imbalance and limited examples of severe artifacts represent the main challenges in this problem.

Our QA approach has several limitations. All DL models employed DenseNet backbones and were trained separately for each task, potentially missing the opportunity to exploit shared representations or multi-task learning. More advanced architectures, such as MambaOut [24], could better integrate the multi-label characteristics of the problem, as our BN approach only models label co-ocurrences without incorporating image features. Additionally, recent approaches based on ordinal contrastive learning have shown promise in handling class imbalance in similar contexts [12,14]. While we considered integrating such methods, time constraints prevented their implementation. Nonetheless, they represent a promising direction for future research.

Our segmentation results for hippocampal segmentation (see Table 2) show that task-specific architectures trained end-to-end on the target dataset (nnU-Net), outperform general purpose foundation models (SAM-Med3D) in uLF settings. This advantage likely stems from nnU-Net's well-established ability to perform reliably with limited data and strong data augmentation strategies. In contrast, SAM-Med3D appears to be limited by domain mismatch, as its pre-training did not include uLF MRI data, highlighting the challenges of transferring knowledge across markedly different imaging domains. For basal ganglia, nnU-Net achieved high accuracy with anatomically consistent boundaries, though performance variability remains high due to the challenging low-contrast nature of uLF MRI. Limitations of this work include the narrow fine-tuning strategy applied to SAM-Med3D, which relied solely on sparse point prompts. More expressive prompting schemes that incorporate anatomical priors or spatial constraints could improve performance. Additionally, exploring alternative PEFT strategies, domain adaptation, and advanced image enhancement techniques such as super-resolution and denoising [11,15,18,20] may further boost segmentation accuracy and robustness in future work.

Acknowledgements. This work was supported by the ERA-NET NEURON MULTI FACT project, funded by the Spanish Ministry of Science, Innovation and Universities (MCIN/AEI/10.13039/501100011033). RGL acknowledges the support of the Generalitat de Catalunya and the European Union through the FI-STEP predoctoral fellowship program (2025 STEP 00603; BDNS 819794).

Disclosure of Interests. The authors declare that there are no conflicts of interest to be disclosed.

References

1. Ankan, A., Textor, J.: pgmpy: a Python toolkit for bayesian networks. J. Mach. Learn. Res. **25**(265), 1–8 (2024). http://jmlr.org/papers/v25/23-0487.html
2. Arnold, T.C., Freeman, C.W., Litt, B., Stein, J.M.: Low-field MRI: clinical promise and challenges. J. Magn. Reson. Imaging **57**(1), 25–44 (2023)

3. Hou, L., Yu, C.P., Samaras, D.: Squared earth mover's distance-based loss for training deep neural networks. arXiv preprint arXiv:1611.05916 (2016)
4. Huang, G., Liu, Z., Van Der Maaten, L., Weinberger, K.Q.: Densely connected convolutional networks. In: Proceedings of the IEEE Conference on Computer Vision and Pattern Recognition, pp. 4700–4708 (2017)
5. Jenkinson, M., Beckmann, C.F., Behrens, T.E., Woolrich, M.W., Smith, S.M.: Fsl. NeuroImage **62**(2), 782–790 (2012). https://doi.org/10.1016/j.neuroimage.2011.09.015
6. Kim, H., Seo, J., Ryu, S., Park, J.H., On, S., Choi, J.: Axis-guided quality assessment and multi-label hippocampal and ventricular segmentation in low-resolution pediatric brain MRI. In: MICCAI Challenge on Low Field Pediatric Brain Magnetic Resonance Image Segmentation and Quality Assurance, pp. 53–62. Springer, Cham (2024)
7. Kingma, D.P., Ba, J.: Adam: a method for stochastic optimization. arXiv preprint arXiv:1412.6980 (2014)
8. Lepore, N., Linguraru, M.G.: Low field pediatric brain magnetic resonance image segmentation and quality assurance: first MICCAI challenge, LISA 2024, Held in Conjunction with MICCAI 2024, Marrakesh, Morocco, 10 October 2024, Proceedings, Springer (2025)
9. Lin, T.Y., Goyal, P., Girshick, R., He, K., Dollár, P.: Focal loss for dense object detection. In: Proceedings of the IEEE International Conference on Computer Vision, pp. 2980–2988 (2017)
10. Liu, Y., et al.: A low-cost and shielding-free ultra-low-field brain MRI scanner. Nat. Commun. **12**(1), 7238 (2021)
11. Lucas, A., et al.: Multi-contrast high-field quality image synthesis for portable low-field MRI using generative adversarial networks and paired data. medRxiv (2023)
12. Mildenberger, D., Hager, P., Rueckert, D., Menten, M.J.: A tale of two classes: adapting supervised contrastive learning to binary imbalanced datasets. In: Proceedings of the Computer Vision and Pattern Recognition Conference, pp. 10305–10314 (2025)
13. Peiris, H., Chen, Z.: Bilateral hippocampi segmentation in low field MRIs using mutual feature learning via dual-views. In: MICCAI Challenge on Low Field Pediatric Brain Magnetic Resonance Image Segmentation and Quality Assurance, pp. 15–27. Springer, Cham (2024)
14. Saleem, A., et al.: SCOL: supervised contrastive ordinal loss for abdominal aortic calcification scoring on vertebral fracture assessment scans. In: International Conference on Medical Image Computing and Computer-Assisted Intervention, pp. 273–283. Springer (2023)
15. Ssentamu, T., et al.: Denoising very low-field magnetic resonance images using native noise modeling. Front. Neuroimaging **4**, 1501801 (2025)
16. Sundaresan, V., Dinsdale, N.K.: Automated quality assessment using appearance-based simulations and hippocampus segmentation on low-field paediatric brain MR images. In: MICCAI Challenge on Low Field Pediatric Brain Magnetic Resonance Image Segmentation and Quality Assurance, pp. 41–52. Springer (2024)
17. Tapp, A., et al.: Quality assurance and hippocampal segmentation on low-field pediatric magnetic resonance images. In: MICCAI Challenge on Low Field Pediatric Brain Magnetic Resonance Image Segmentation and Quality Assurance, pp. 63–75. Springer, Cham (2024)
18. Tapp, A., et al.: Super-field MRI synthesis for infant brains enhanced by dual channel latent diffusion. In: International Conference on Medical Image Computing and Computer-Assisted Intervention, pp. 444–454. Springer (2024)

19. Torrents-Barrena, J., Piella, G., Masoller, N., Gratacós, E., Eixarch, E., Ceresa, M., González Ballester, M.A.: Segmentation and classification in MRI and US fetal imaging: recent trends and future prospects. Med. Image Anal. **51**, 61–88 (2019)
20. Vega, F., Addeh, A., MacDonald, M.E.: Denoising simulated low-field MRI (70mT) using denoising autoencoders (DAE) and cycle-consistent generative adversarial networks (cycle-GAN). arXiv preprint arXiv:2307.06338 (2023)
21. Wang, H., et al.: SAM-Med3D: towards general-purpose segmentation models for volumetric medical images. In: European Conference on Computer Vision, pp. 51–67. Springer (2024)
22. Wang, X., Lyul, M.: Infant hippocampal segmentation in ultra-low-Field MRI using external. Low Field Pediatric Brain Magnetic Resonance Image Segmentation and Quality Assurance: First MICCAI Challenge, LISA 2024, Held in Conjunction with MICCAI 2024, Marrakesh, Morocco, 10 October 2024, Proceedings **15515**, 28 (2025)
23. Zhu, Y., Jiang, H., Cai, R., Chen, G.: Multi-label MambaOut for quality assessment of low-field pediatric brain MR images. In: MICCAI Challenge on Low Field Pediatric Brain Magnetic Resonance Image Segmentation and Quality Assurance, pp. 3–11. Springer, Cham (2024)
24. Zhu, Y., Jiang, H., Cai, R., Chen, G.: Multi-label MambaOut for quality assessment of low-field pediatric brain MR images. In: Lepore, N., Linguraru, M.G. (eds.) Low Field Pediatric Brain Magnetic Resonance Image Segmentation and Quality Assurance, pp. 3–11. Springer Nature Switzerland, Cham (2025)

Correction to: Automatic Quality Assurance and Subcortical Brain Segmentation in Pediatric Ultra-Low-Field MRI: Exploring Ordinal Learning and Foundation Model Adaptation

Raquel González López, Maria Chiara Fiorentino, Gerard Martí-Juan, Oscar Camara, and Miguel A. González Ballester

Correction to:
Chapter 11 in: N. Lepore and M. G. Linguraru (Eds.):
Low Field Pediatric Brain Magnetic Resonance Image
Segmentation and Quality Assurance, **LNCS 16411,**
https://doi.org/10.1007/978-3-032-14417-1_11

The book was published with a typo of chapter 11 in this book ID (686810_1_En). The final published version did not include the acknowledgments section. This section has now been added to include the following: "This work was supported by the ERA-NET NEURON MULTI FACT project, funded by the Spanish Ministry of Science, Innovation and Universities (MCIN/AEI/10.13039/501100011033). RGL acknowledges the support of the Generalitat de Catalunya and the European Union through the FI-STEP predoctoral fellowship program (2025 STEP 00603; BDNS 819794)."

© The Author(s) 2026
N. Lepore and M. G. Linguraru (Eds.): LISA 2025, LNCS 16411, p. C1, 2026.
https://doi.org/10.1007/978-3-032-14417-1_12

Author Index